Advanced Optimization and Decision-Making Techniques in Textile Manufacturing

T0136214

Advanced Optimization and Decision-Making Techniques in Textile Manufacturing

Anindya Ghosh
Prithwiraj Mal
Abhijit Majumdar

CRC Press
Taylor & Francis Group
Boca Raton London New York

CRC Press is an imprint of the
Taylor & Francis Group, an **informa** business

MATLAB® is a trademark of The MathWorks, Inc. and is used with permission. The MathWorks does not warrant the accuracy of the text or exercises in this book. This book's use or discussion of MATLAB® software or related products does not constitute endorsement or sponsorship by The MathWorks of a particular pedagogical approach or particular use of the MATLAB® software.

CRC Press
Taylor & Francis Group
6000 Broken Sound Parkway NW, Suite 300
Boca Raton, FL 33487-2742

First issued in paperback 2020

© 2019 by Taylor & Francis Group, LLC
CRC Press is an imprint of Taylor & Francis Group, an Informa business

No claim to original U.S. Government works

ISBN-13: 978-1-138-58676-5 (hbk)
ISBN-13: 978-0-367-67053-5 (pbk)

Visit the Taylor & Francis Web site at
http://www.taylorandfrancis.com

and the CRC Press Web site at
http://www.crcpress.com

Contents

Preface

Optimization is the selection of the overall best course of action or alternative from a set of available options. Decision-making is a cognitive process that is often based on optimization. Optimization and decision-making are an integral part of any manufacturing process and management system. The objective of this text is to demonstrate the confluence of theory and applications of various types of multicriteria decision-making and optimization techniques with reference to textile manufacturing and management. At the beginning, multicriteria decision-making and optimization techniques are discussed at a basic level. The concepts and algorithms are explained in a step-by-step method so that readers can easily comprehend and solve various textile problems involving optimization and decision-making. An elementary knowledge of school-level mathematics would be sufficient for understanding most of the concepts discussed here. The authors have provided adequate examples and case studies related to textile manufacturing processes for each concept discussed in each chapter. For better understanding, readers may utilize the MATLAB® coding provided with each chapter to solve various types of multicriteria decision-making and optimization problems.

This text is divided into 12 chapters. An introduction on the importance of multicriteria decision-making and optimization techniques has been provided. Various multicriteria decision-making methods such as analytic hierarchy process (AHP), fuzzy AHP, technique for order preference by similarity to ideal solution (TOPSIS), fuzzy TOPSIS, elimination and choice expressing reality (ELECTRE), and graph theory and matrix approaches have been discussed with suitable examples. Subsequently, different types of traditional and nontraditional optimization techniques like linear programming, fuzzy linear programming, quadratic programming, genetic algorithm, particle swarm optimization, and simulated annealing have been explained with applications in the textile domain. Finally, multiobjective optimization problems have been dealt with different approaches, such as goal programming, desirability function, and evolutionary algorithm. A thorough literature survey of each concept has also been included.

This text is expected to become essential input for the textile industry as well as academia, as it will help to present optimization and decision-making in an efficient and scientific way.

Anindya Ghosh
Prithwiraj Mal
Abhijit Majumdar

MATLAB® is a registered trademark of The MathWorks, Inc. For product information, please contact:

The MathWorks, Inc.
3 Apple Hill Drive
Natick, MA 01760-2098 USA
Tel: 508 647 7000
Fax: 508-647-7001
E-mail: info@mathworks.com
Web: www.mathworks.com

About the Authors

Anindya Ghosh completed his BTech in Textile Technology in 1997 from the College of Textile Technology, Berhampore (Calcutta University), India. After that he worked in the textile spinning industry for 1 year as a shift engineer. He completed his MTech and PhD degrees in Textile Engineering from the Indian Institute of Technology, Delhi, India, in 2000 and 2004, respectively. He is a recipient of the Career Award for Young Teacher – 2009 from AICTE, India. He has more than 14 years of experience in teaching. Currently, he is working as an associate professor at the Government College of Engineering and Textile Technology, Berhampore. His research work involves yarn manufacturing, yarn and fabric structures, modeling and simulation, optimization, and decision-making techniques. He has published more than 70 papers in various refereed journals.

Prithwiraj Mal is a postgraduate in Textile Engineering at the Indian Institute of Technology, Delhi, India. He has almost 16 years of cumulative experience in both industry and academics. He completed a PhD degree at Jadavpur University, Kolkata, in 2017. Prithwiraj Mal joined the National Institute of Fashion Technology (NIFT) in 2008 and currently posted in NIFT Hyderabad as an assistant professor in the Department of Textile Design. His research work involves comfort, optimization, decision-making techniques, and product development. He has published more than 20 papers in various refereed journals and presented and published papers at conferences at national and international levels.

 **Abhijit Majumdar** is working as a professor in the Department of Textile Technology at the Indian Institute of Technology Delhi, India. A graduate from Calcutta University, with a gold medal in Textile Technology, he completed his postgraduate and PhD degrees from the Indian Institute of Technology (IIT) Delhi and Jadavpore University, Kolkata, respectively. His research areas include protective textiles (soft armor, antibacterial, ultraviolet, etc.), soft computing applications, and sustainable supply chain management. He completed research projects funded by the Department of Science and Technology (DST), Defense Research and Development Organization (DRDO), and Council for Scientific and Industrial Research (CSIR). He has published 85 research papers in international refereed journals and guided seven PhD students. He authored one book entitled *Principles of Woven Fabric Manufacturing*, published by CRC Press. He also edited two books published by Woodhead Publisher, United Kingdom, and authored one monograph (*Textile Progress*) published by Taylor and Francis Group. He is the associate editor of the *Journal of the Institution of Engineers* (India), Series E (Chemical and Textile Engineering), published by Springer. He is a recipient of the Outstanding Young Faculty Fellowship (2009–2014) of IIT Delhi, Teaching Excellence Award (2015) of IIT Delhi, and Gandhian Young Technological Innovation Award (2017).

1

Introduction to Decision-Making and Optimization Techniques

1.1 Introduction

Decision-making is the art and science of choosing the best course of action or a set of preferred actions from the available alternatives. Decision-making involves assumptions, information, emotion, communication, and computation. In an organization, decisions are taken at three levels, namely, operational, tactical, and strategic. Operational decisions are taken at the lower-level management based on structured information, and these decisions have short-term impact. Strategic decisions, which involve unstructured information, are taken by the top management, and they have long-term implications. Tactical decisions are taken by the middle management. All of these decisions are taken after considering information related to political, economic, social, and technological (PEST) environments.

Optimization, which is often used as a tool for decision-making, is one of the most exciting branches of applied mathematics. There is hardly any area in science and engineering where optimization has not made itself useful. The great scientist E. Euler once quoted, "Nothing takes place in the world whose meaning is not that of some maximum or minimum." With the advent of high-speed computers, optimization algorithms are becoming increasingly popular in science and engineering. Over the last few decades, the introduction of a number of nontraditional optimization algorithms inspired by natural phenomena has added a new dimension to the field of optimization.

1.2 Decision-Making Process and Classification

The process of decision-making is shown in Figure 1.1. It is a sequential process, and the individual steps should not be considered in isolation. Decision-making becomes simple if only one alternative is available as there is

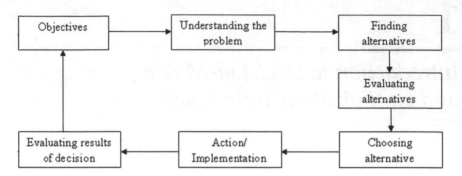

FIGURE 1.1
Decision-making process.

nothing to choose, or if one alternative dominates all others in every criterion. However, in real life this kind of scenario is seldom obtained.

Decision-making can broadly be classified under three heads, namely, traditional, multiattribute, and fuzzy, as depicted in Figure 1.2. Traditional decision-making varies based on the knowledge about future conditions. Future conditions associated with decision-making vary along a continuum ranging from complete certainty to complete uncertainty. Decision-making under certainty, under risk, and under uncertainty are the variants of traditional decision-making. In case of decision-making under certainty, the

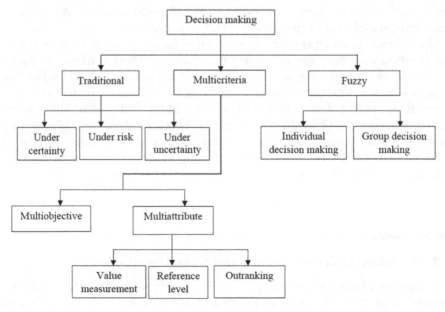

FIGURE 1.2
Classification of decision-making techniques.

decision maker has complete and accurate information about the outcome of each of the alternative actions (decision choices). Break-even analysis, linear programming, etc., are examples of decision-making under certainty. In case of decision-making under risk, the decision maker is aware of the possible future states of nature and their probabilities. Using a decision tree is one method for making decisions under risk. In case of decision-making under uncertainty, the decision maker is unaware of the probabilities of future states of nature that are associated with alternative actions.

Multicriteria decision-making (MCDM) can involve either multiple objectives to be fulfilled simultaneously or multiple attributes on which the alternatives are evaluated. MCDM is a full-grown branch of operations research (OR) that, in recent years, has generated a lot of interest among theoretical and applied researchers. MCDM problems can be classified into multiobjective or multiattribute based on the nature (design or selection) of the problem. Multiattribute decision-making (MADM) can have different philosophies based on utility value measurement, comparison with the reference levels, and outranking. Analytic hierarchy process (AHP), technique for order preference by similarity to ideal solutions (TOPSIS), and elimination and choice expressing reality (ELECTRE) are examples of MADM techniques working with value measurement, reference levels, and outranking approaches, respectively.

Fuzzy decision-making involves the handling of imprecise information. This may further be classified under individual decision-making and group decision-making. Fuzzy linear programming, fuzzy MCDM, etc., are examples of fuzzy decision-making.

1.2.1 Decision-Making Under Certainty, Risk, and Uncertainty

Table 1.1 presents the payoff matrix for a make or buy decision. The payoff has been indicated in terms of net profit for the three courses of action. In case of decision-making under certainty, the decision maker would know the state of nature that will prevail in the future with complete certainty. Therefore, he or she will use only one part of Table 1.1 for decision-making. For example, if the decision maker is certain that demand for the product will be low, the best decision would be to "buy gray fabric," as it will give a higher net profit of $6 million.

TABLE 1.1

Payoff Table (Net Profit) for Decision-Making

Actions	States of Nature (Demand for Finished Fabric)		
	(Low)	(Moderate)	(High)
Make gray fabric	$5 million	$8 million	$15 million
Buy gray fabric	$6 million	$8 million	$10 million
Make and buy gray fabric	$2 million	$4 million	$20 million

TABLE 1.2

Payoff Table for Decision-Making Under Risk

Actions	States of Nature (Demand for Finished Fabric)		
	(Low, 0.5)	(Moderate, 0.3)	(High, 0.2)
Make gray fabric	$2.5 million	$2.4 million	$3 million
Buy gray fabric	$3 million	$2.4 million	$2 million
Make and buy gray fabric	$1 million	$1.2 million	$4 million

Table 1.2 presents the payoff table for decision-making under risk. In this case, the decision maker is aware of the probabilities associated with all possible future states of nature as shown in Table 1.2. The net profit values shown in Table 1.1 have been multiplied by the corresponding probabilities to get the required net profit values. Now, the expected net profit for "make gray fabric" is $7.9 million, whereas for "buy gray fabric" it is $7.4 million, and for "make and buy gray fabric," it is $6.2 million. Therefore, the decision maker would go with "make gray fabric."

In case of decision-making under uncertainty, the decision maker does not have any knowledge or information about the chances of future states of nature. Under such a scenario, the decisions can be taken based on the following criteria:

- Maximax criterion: Maximizing the maximum payoff
- Maximin criterion: Maximizing the minimum payoff
- Minimax criterion: Minimizing the maximum regret
- Laplace criterion
- Hurwicz criterion

Maximax Criterion

The maximax criterion is used by the most optimistic decision maker. It is assumed that the most favorable condition will prevail in the future. From Table 1.1, the most favorable and unfavorable payoffs for the three alternative actions have been calculated and are shown in Table 1.3. As the objective is to

TABLE 1.3

Payoff Table for Decision-Making Under Uncertainty

Actions	States of Nature (Demand for Finished Fabric)			Maximum Payoff	Minimum Payoff
	(Low)	(Moderate)	(High)		
Make gray fabric	$5 million	$8 million	$15 million	$15 million	$5 million
Buy gray fabric	$6 million	$8 million	$10 million	$10 million	$6 million
Make and buy gray fabric	$2 million	$4 million	$20 million	$20 million	$2 million

maximize, the decision would be to "make and buy gray fabric." The decision based on the maximax criterion may backfire as there is no guarantee that the most favorable state of nature will prevail in the future.

Maximin Criterion

This criterion is used by the pessimistic decision maker. It is assumed that the most unfavorable state of nature will prevail in the future. From Table 1.3, it is observed that minimum payoffs are $5, $6, and $2 million for make, buy, and make and buy decisions, respectively. As the objective is to maximize, the decision would be to "buy gray fabric."

Minimax Criterion

The minimax criterion is used by the decision maker to minimize the maximum regret. When the decision maker chooses an action, he or she may regret if the chosen action does not yield maximum payoff. Regret is defined as the difference in payoff between the best and chosen actions under a particular state of nature. In other words, regret is the opportunity cost for not choosing the best action for a particular state of nature. From Table 1.1, the regret Table 1.4 has been prepared. The maximum regret is $5, $10, and $4 million for make, buy, and make and buy actions, respectively. As the objective is to minimize the maximum regret, the decision would be to "make and buy gray fabric."

Laplace Criterion

The Laplace criterion is used when there is no knowledge about the probabilities of future states of nature. For simplicity, it may be assumed that all of them are equally likely. Table 1.5 shows the payoff considering one-third probability for all three actions. Therefore, the expected payoff for make, buy, and make and buy is $9.34, $8, and $8.67 million, respectively. As the payoff is maximum for "make gray fabric," it should be the chosen decision as per the Laplace criterion.

TABLE 1.4

Regret Table for Decision-Making Under Uncertainty

| Actions | States of Nature (Demand for Finished Fabric) | | | Maximum Regret |
	(Low)	(Moderate)	(High)	
Make gray fabric	$1 million	0	$5 million	$5 million
Buy gray fabric	0	0	$10 million	$10 million
Make and buy gray fabric	$4 million	$4 million	0	$4 million

TABLE 1.5

Payoff Table for Laplace Criterion

Actions	States of Nature (Demand for Finished Fabric)		
	(Low, 1/3)	(Moderate, 1/3)	(High, 1/3)
Make gray fabric	$1.67 million	$2.67 million	$5 million
Buy gray fabric	$2 million	$2.67 million	$3.33 million
Make and buy gray fabric	$0.67 million	$1.33 million	$6.67 million

TABLE 1.6

Weighted Payoff Based on Degrees of Optimism and Pessimism

Actions	Maximum Payoff ($\alpha = 0.7$)	Minimum Payoff $(1 - \alpha = 0.3)$	Weighted Payoff
Make gray fabric	$15 million	$5 million	$12 million
Buy gray fabric	$10 million	$6 million	$8.8 million
Make and buy gray fabric	$20 million	$2 million	$14.6 million

Hurwicz Criterion

The maximax and maximin criteria discussed are far from realistic decisions, as the decision maker is never completely optimistic or completely pessimistic. If the degree of optimism is α, then the degree of pessimism is $(1 - \alpha)$. Using the degrees of optimism and pessimism as 0.7 and 0.3, respectively, a weighted payoff is calculated as shown in Table 1.6. As the weighted payoff is maximum ($14.6 million) for "make and buy gray fabric," it should be chosen according to the Hurwicz criterion.

1.2.2 Multicriteria Decision-Making

Multicriteria decision-making (MCDM) deals with making decisions in the presence of multiple criteria that are usually conflicting (price and quality). The problems of MCDM can be broadly classified under two heads, namely, multiattribute decision-making (MADM) and multiobjective decision-making (MODM). The former is used in case of a selection problem, whereas the latter is in case of a design problem. The salient differences between the two methods have been presented in Table 1.7.

MADM deals with the selection of the best alternative or a set of preferred alternatives under the presence of a finite number of decision criteria and alternatives. Weighted-sum model (WSM), weighted-product model (WPM), AHP model, TOPSIS, ELECTRE model, decision-making trial and evaluation laboratory (DEMATEL) technique, and preference ranking organization method for enrichment of evaluations (PROMETHEE) are some of the widely used methods of MCDM.

TABLE 1.7

Differences between MADM and MODM Methods

MADM Method	MODM Methods
Decision variables are generally discrete.	Decision variable values are determined in a continuous or integer domain.
Number of alternatives are finite.	The number of alternatives are infinite when variables are continuous and very large when variables are integers.
Alternatives are predetermined.	Alternatives are not predetermined and are obtained by solving mathematical problems.

The major steps of MADM are as follows:

- Determining the objective or goal, relevant criteria, and alternatives of the problem
- Determining the relative importance (weights) of criteria with respect to the objective
- Determining the scores of all the alternatives with respect to each of the decision criteria
- Processing the criteria weights and alternative scores to determine the priority score and ranking of each alternative

The WSM, proposed by Fishburn (1967), is probably the most popularly used approach in single-dimensional cases. According to this model, the best alternative is the one that satisfies the following expression in case of maximization:

$$A_{WSM} = \max \sum_{j=1}^{N} a_{ij} W_j, \quad \text{for } i = 1, 2, 3, \dots, M. \quad (1.1)$$

where A_{WSM} is the score of the best alternative, N is the number of decision criteria, M is the number of alternatives, W_j is the weight of the jth criterion, and a_{ij} is the actual value of the ith alternative in terms of the jth criterion.

The WSM performs well in single-dimensional cases, where all the scores, given to alternatives, are having the same unit. However, it is not acceptable in case of multidimensional problems, which involve different units for different criteria. In WPM, each alternative is compared with the others by multiplying a number of ratios, one for each criterion. The WPM is also known as dimensionless analysis as it eliminates the units of various measures for criteria. Therefore, it can be used for single as well as multidimensional problems. Bridgman (1922) and Miller and Starr (1969) proposed the following expression to compare the alternatives A_K and A_L:

$$R(A_K / A_L) = \prod_{j=1}^{N} (a_{Kj} / a_{Lj})^{W_j} \quad (1.2)$$

where a_{Kj} and a_{Lj} are the scores received by the alternatives A_K and A_L, respectively, in the jth criterion. If the term $R(A_K/A_L)$ is greater than one, then the alternative A_K is more desirable than A_L, and vice versa.

AHP and TOPSIS have been popularly used in many decision-making problems related to the textile industry, such as cotton fiber grading (Majumdar et al., 2005), naval selection for rotor spinning (Majumdar et al., 2010), jute fiber grading (Ghosh and Das, 2013), and ranking of handloom fabrics (Mitra et al., 2015). Therefore, these two methods have been dealt with in detail in subsequent chapters of this text.

1.3 Optimization

Optimization is the process of obtaining the best possible result out of all the conditions and available means. There is hardly any area in science and engineering where optimization has not made itself useful. Thus, optimization has become one of the most exciting branches of applied mathematics. An optimization problem either minimizes or maximizes a mathematical function, involving a set of variables, subject to a set of constraints. The mathematical function to be minimized or maximized is known as *objective function*. The other conditions to be fulfilled are termed as *constraints*. If the objective function as well as the constraints are linear functions of variables, then the problem is called a *linear optimization problem*. If the objective function or any of the constraint equations involves nonlinearity, then it is classified as *nonlinear optimization*. A classification of optimization problem is shown in Figure 1.3 (Majumdar et al., 2011). Several nontraditional optimization techniques,

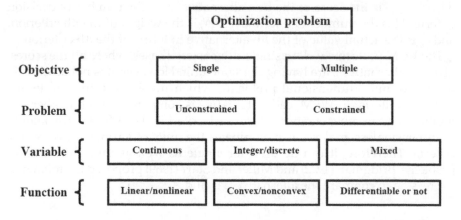

FIGURE 1.3
Classification of optimization problem. (From Majumdar, A. et al. 2011. *Indian Journal of Fibre and Textile Research*, 36, 398–409.)

like genetic algorithm, particle swarm optimization, simulated annealing, etc., became popular in the recent past. Some traditional and nontraditional optimization techniques are introduced in the following sections.

1.3.1 Linear Programming

Linear programming is the simplest optimization technique that attempts to maximize or minimize a linear function of decision variables. The values of the decision variables are chosen such that a set of restricting conditions is satisfied. Linear programming involving only two decision variables can be solved by using a graphical method. However, an iterative simplex method is used to solve linear programming problems involving three or more decision variables. Linear programming is very commonly used to solve the product mix problem of manufacturing industries. An example has been presented here for the understanding of the readers.

Let two sizes of functional clothing, namely, M and L, be manufactured in an industry that aims at maximization of overall profit. Profit per unit sales is Rs. 5,000 and 10,000 for sizes M and L, respectively. The machine hour requirements per unit production are 2 and 2.5 for sizes M and L, respectively. The company must produce at least 10 functional clothing pieces in a day to meet the market demand. The stated facts can be converted to a linear programming problem as follows:

$$\text{Objective function:} \quad \text{Maximize:} \, 5,000M + 10,000L \tag{1.3}$$

$$\text{Subject to:} \quad 2M + 2.5L \leq 24 \tag{1.4}$$
$$M + L \geq 10$$

After solving the above linear programming problem, it is found that the maximum profit of the industry will be Rs. 90,000 per day provided it manufactures 2 and 8 units of functional clothing of sizes M and L, respectively. A graphical representation of this linear programming problem is depicted in Figure 1.4.

Linear programming cannot be used when the objective and constraint functions are nonlinear in nature. In such cases, nonlinear programming techniques can be used to solve the optimization problem. The optimization literature contains a large number of nonlinear programming techniques. A quadratic programming process is one of the nonlinear programming techniques that is used when the problem has a nonlinear quadratic objective function and linear constraint functions.

1.3.2 Multiobjective Optimization and Goal Programming

Adding multiple objectives to an optimization problem increases the computational complexity. For example, if the design of ultraviolet protective

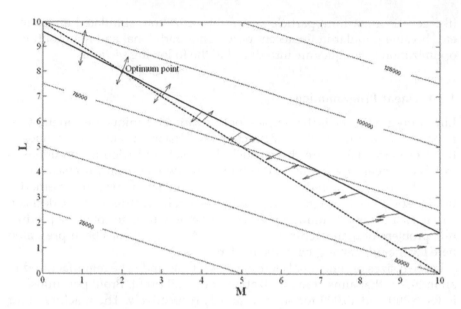

FIGURE 1.4
Optimum point of linear programming problem. (From Majumdar, A. et al. 2011. *Indian Journal of Fibre and Textile Research*, 36, 398–409.)

clothing has to be optimized to provide good air permeability, then these two objectives conflict and a trade-off is needed. There will be one design that will provide maximum ultraviolet protection factor (UPF) but minimum air permeability. But, there will be another design that will provide minimum UPF but maximum air permeability. Between these two extreme designs, an infinite number of designs will exist which are of some compromise between UPF and air permeability. This set of trade-off designs is known as a Pareto set. The curve created by plotting objective one (UPF) against objective two (air permeability) for the best designs is known as the Pareto frontier. None of the solutions in the Pareto front is better than the other—that is, any one of them is an acceptable solution. The choice of one design solution over another depends exclusively on the requirement of the process engineer. Majumdar et al. (2011) developed the Pareto optimal front for UPF and air permeability of cotton woven fabrics as depicted in Figure 1.5. The optimal design fronts are different for various yarn linear densities. It is observed that for a fabric having a UPF value of 30, the air permeability will be better if it is woven using 20 Ne weft yarns.

The goal programming technique is often used to solve multiobjective optimization problems. In goal programming, a numeric goal is established for each goal function or constraint. The objective function minimizes the weighted sum of undesirable deviations from the respective goals. The example given in the previous section can be converted to a goal programming problem assuming that the profit goal of the organization is Rs. 90,000:

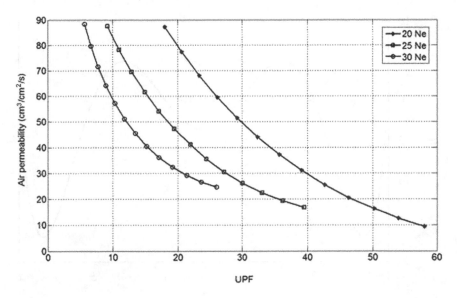

FIGURE 1.5
Pareto optimal front for UPF and air permeability. (From Majumdar, A. et al. 2011. *Indian Journal of Fibre and Textile Research*, 36, 398–409.)

$$5,000M + 10,000L + d_1^- - d_1^+ = 90,000$$
$$2M + 2.5L + d_2^- - d_2^+ = 24 \tag{1.5}$$
$$M + L + d_3^- - d_3^+ = 10$$

$$\text{Minimize} = w_1 d_1^- + w_2 d_2^+ + w_3 d_3^- \tag{1.6}$$

Here d_1^- and d_1^+ are deviational variables for machine hour, d2- and d2+ are deviational variables for number of functional clothing, w_1, w_2, and w_3 are the weights assigned to the deviational variables.

1.3.3 Nontraditional Optimization Algorithms

Nontraditional algorithms are attempts to mimic natural phenomena to solve complex engineering problems. A glimpse of some widely established nontraditional algorithms such as the genetic algorithm, particle swarm optimization, and simulated annealing are given in the following sections.

1.3.3.1 Genetic Algorithm

The genetic algorithm (GA) is an unconventional search method based on a natural selection process for solving complex optimization problems. It was developed by John Holland (1975) of the University of Michigan. Unlike

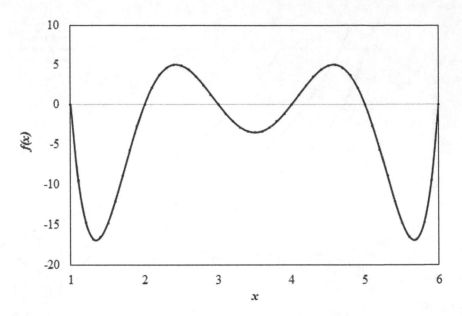

FIGURE 1.6
Function having local and global minima. (From Majumdar, A. et al. 2011. *Indian Journal of Fibre and Textile Research*, 36, 398–409.)

conventional derivative-based optimization that requires differentiability of the function to be optimized, GA can handle functions with discontinuities or piecewise segments. Gradient-based optimization algorithms can get stuck in local minima or maxima as they rely on the slope of the function. A GA overcomes this problem. The following function has local and global minima as shown in Figure 1.6:

$$f(x) = (x-1)(x-2)(x-3)(x-4)(x-5)(x-6) \qquad (1.7)$$

Gradient-based optimization, while searching for the global minima, may get stuck at 3.5, which is actually a local minima. However, GA is certain to find out the global minima of the function at 1.34.

To perform the optimization task, GA maintains a population of points called "individuals," each of which is a potential solution to the optimization problem. Generally the individuals are coded with a string of binary numbers. The GA repeatedly modifies the population of individual solutions using selection, crossover, and mutation operators. At each step, the GA selects individuals from the current population (parents) and uses them to produce children for the next generation, which competes for survival. Over successive generations, the population "evolves" toward an optimal solution. GA can be applied to solve a variety of optimization problems where the objective function is discontinuous, nondifferentiable, stochastic, or highly nonlinear.

1.3.3.2 Particle Swarm Optimization

Particle swarm optimization (PSO) is a population-based stochastic optimization technique motivated by the social behavior of bird flocking. It mimics the choreography of a bird swarm that flies through the problem space, where each bird adjusts its traveling speed dynamically corresponding to the flying experiences of itself and its colleagues (Kennedy and Eberhart, 1995). The PSO uses a number of particles that represent a bird swarm flying within the search space looking for the best solution. It begins with a population of solutions where each solution is regarded as the position of a particle in the search space. All of the particles in a swarm have fitness values that can be calculated from the objective function. Each particle in the swarm preserves its individual best performance, and in addition, it keeps track of the best performance of the swarm. Each particle continually adjusts its position by modifying the velocity according to its individual best performance as well as the best performance of the swarm. Sooner or later, most of the particles converge to the optimum solution.

1.3.3.3 Simulated Annealing

SA is a useful meta-heuristic for solving hard combinatorial optimization problems. It was first introduced by Kirkpatrick et al. (1983). It simulates the cooling phenomenon of molten polymers or metals to form a search and optimization procedure. SA is a step-by-step method that could be considered as an improvement of the local optimization algorithm. The local optimization algorithm proceeds by generating, at each iteration, a solution in the neighborhood of the previous one. If the value of a criterion corresponding to the new solution is better than the previous one, the new solution is selected, otherwise it is rejected. The SA algorithm terminates either when it is no longer possible to improve the solution or the maximum number of trials decided by the user is reached. The main drawback of the local optimization algorithm is that it terminates at a local minimum, which depends on the initial solution and may be far from the global minimum.

SA begins with a randomly selected current point in the search space, and the initial temperature T is kept at high value. A new point is generated at random in the neighborhood of the current point. Then the difference in the function values ΔE at these two points is measured, and the Metropolis algorithm is applied in order to accept or reject the point. If the new point is accepted, it is taken as a starting point for the next step. However, if the new point is rejected, the original point is retained as a starting point for the next step. In the next iteration, again a new point is randomly generated in the vicinity of the current point, and the Metropolis algorithm is used to accept or reject the new point. In general, before reducing the temperature, a number of iterations are carried out at a particular temperature. The temperature is lowered according to the cooling schedule, and cooling is done at a slower rate. The algorithm is stopped when temperature becomes just lower than the termination temperature.

1.4 Summary

A brief discussion of decision-making and optimization techniques has been presented in this chapter. The basics of the decision-making process such as its classification, decision-making under certainty, risk and uncertainty, and multicriteria decision-making have been discussed. An account of traditional and nontraditional optimization techniques is also given. The various techniques discussed in this chapter provide the background of the applications of decision-making and optimization in the domain of textile engineering that are discussed in the rest of this book.

References

Bridgman, P. W. 1922. *Dimensional Analysis*, Yale University Press, New Haven, CT.

Fishburn, P. C. 1967. *Additive Utilities with Incomplete Product Set: Applications to Priorities and Assignments*, American Society of Operations Research (ORSA), Baltimore, MD.

Ghosh, A. and Das, S. 2013. Raw jute grading by multi-criteria decision making method. *Journal of Natural Fibers*, 10, 136–146.

Holland, J. H. 1975. *Adaptation in Natural and Artificial System*, University of Michigan Press, Ann Arbor, MI.

Kennedy, J. and Eberhart, R. C. 1995. Particle swarm optimization. *Proceedings of IEEE International Conference on Neural Networks*, Perth, Australia, 1942–1948.

Kirkpatrick, S., Gelatt, C. D. and Vecchi, M. P. 1983. Optimization by simulated annealing. *Science*, 220, 671–680.

Majumdar, A., Kaplan, S. and Göktepe, Ö. 2010. Navel selection of rotor spinning for denim fabric by using multi-criteria decision making process. *Journal of the Textile Institute*, 101, 304–309.

Majumdar, A., Majumdar, P. K. and Sarkar, B. 2005. Determination of technological value of cotton fibre: A comparative study between traditional and multiple criteria decision making approach. *Autex Research Journal*, 5, 71–80.

Majumdar, A., Singh, S. P. and Ghosh, A. 2011. Modelling, optimization and decision making techniques in designing of functional clothing. *Indian Journal of Fibre and Textile Research*, 36, 398–409.

Miller, D. W. and Starr, M. K. 1969. *Executive Decisions and Operations Research*, Prentice Hall, Englewood Cliffs, NJ.

Mitra, A., Majumdar, A., Ghosh, A., Majumdar, P. K. and Bannerjee, D. 2015. Selection of handloom fabrics for summer clothing using multi-criteria decision making techniques. *Journal of Natural Fibers*, 12, 61–71.

2

Analytic Hierarchy Process

2.1 Introduction

The analytic hierarchy process (AHP) is one of the most popular multi-criteria decision-making (MCDM) methods that deals with selection of the best alternative or ranking of alternatives under the presence of a finite number of decision criteria. In addition to AHP, the weighted-sum model (WSM), the weighted-product model (WPM), a technique for order preference by similarity to ideal solutions (TOPSIS), the elimination and choice translating reality (ELECTRE), the decision-making trial and evaluation laboratory (DEMATEL), etc., are some of the widely used exponents of MCDM.

AHP was invented by Saaty (1980, 1983, 1986), and the first two papers entitled "Estimating technological coefficients by the analytic hierarchy process" and "The U.S.-OPEC energy conflict the payoff matrix by the Analytic Hierarchy Process," were published in *Socio-Economic Planning Sciences* and in the *International Journal of Game Theory*, respectively, in the year 1979. A search in the Scopus database shows an exponential rise in the use of AHP in the last four decades (Figure 2.1). Although there are some issues and challenges with AHP (Belton and Gear, 1983; Saaty and Vargas, 1984; Dyer, 1990a,b; Harker and Vargas, 1990; Triantaphyllou and Mann, 1994; Triantaphyllou, 2001), it has proven to be a simple and extremely useful method of MCDM. Applications of AHP have been found in almost every domain of engineering, management, and social sciences, including a ranking of livable cities; strategic decisions like facility location, merger and acquisition, research and development project selection, etc.; information technology outsourcing decisions; operational decisions like software selection, supplier selection, maintenance, logistics etc.; benchmarking; and engineering education.

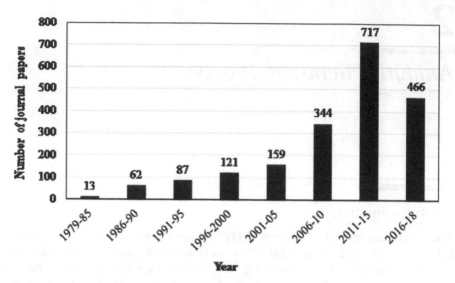

FIGURE 2.1
Number of journal papers with "Analytic Hierarchy Process" in title.

2.2 Analytic Hierarchy Process Methodology

Step 1

First, the hierarchical structure of the problem is formulated. The overall objective or goal of the decision problem and alternatives are placed at the top and bottom of the hierarchy, respectively. Relevant criteria and subcriteria are positioned at the intermediate levels as shown in Figure 2.2. The number of levels in the hierarchy depends on the complexity of the problem.

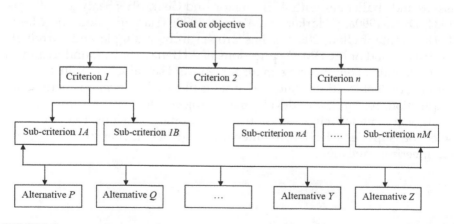

FIGURE 2.2
Structural hierarchy of AHP.

Step 2

The relational data to determine the weightage of criteria and subcriteria with respect to the objective of the problem are developed in this step. This requires the decision maker to formulate pair-wise comparison matrices of elements at each level of the hierarchy relative to the corresponding element at the next higher level. In AHP, if a problem involves M alternatives and N criteria, then the decision maker has to construct N judgment matrices of alternatives of $M \times M$ order and one judgment matrix of criteria of $N \times N$ order. Finally, the decision matrix of $M \times N$ order is formed by using the relative scores of the alternatives with respect to each criterion. In AHP, a relational scale of real numbers from 1 to 9 and their reciprocals are used to assign preferences in a systematic manner. When comparing two criteria (or alternatives) with respect to an attribute at a higher level, the relational scale proposed by Saaty (1980, 1983, 1986) is used. The scale is shown in Table 2.1.

Step 3

In this step, the relative weightage or importance of different criteria with respect to the goal of the problem and the alternative scores with respect to each criterion are determined. For N criteria, the size of the pair-wise comparison matrix (C_1) will be $N \times N$, and the entry c_{ij} will denote the relative

TABLE 2.1

The Fundamental Relational Scale for Pair-Wise Comparisons

Intensity of Importance on an Absolute Scale	Definition	Explanation
1	Equal importance	Two activities contribute equally to the objective.
3	Moderate importance of one over another	Experience and judgment slightly favor one activity over another.
5	Essential or strong importance	Experience and judgment strongly favor one activity over another.
7	Very strong importance	An activity is strongly favored, and its dominance is demonstrated in practice.
9	Extreme importance	The evidence favoring one activity over another is of the highest possible order of affirmation.
2, 4, 6, 8	Intermediate values between two adjacent judgment	When compromise is needed.
Reciprocals	If activity p has one of the above numbers assigned to it when compared with activity q, then q has the reciprocal value when compared with p.	

importance of criterion i with respect to the criterion j. In the matrix, $c_{ij} = 1$ when $i = j$ and $c_{ij} = (1/c_{ji})$:

$$C_1 = \begin{bmatrix} 1 & c_{12} & \cdots & c_{1N} \\ c_{21} & 1 & \cdots & c_{2N} \\ \cdots & \cdots & 1 & \cdots \\ c_{N1} & c_{N2} & \cdots & 1 \end{bmatrix}$$

The relative weight or importance of the ith criterion (W_i) is determined in two steps: First, normalize each entry of the matrix so that the sum of each column becomes 1, in other words, all the entries of the matrix are divided by the sum of a corresponding column. Second, calculate the row averages to get the weight of different elements (criteria). Alternatively, the geometric mean (GM) of the rows of the matrix C_1 is calculated and then they are normalized, as shown in the following equation, to elicit the weight of ith criteria (W_i):

$$GM_i = \left\{ \prod_{j=1}^{N} c_{ij} \right\}^{1/N} \quad \text{and} \quad W_i = \frac{GM_i}{\sum_{i=1}^{N} GM_i} \tag{2.1}$$

The following example of selection of the best fabric will make it clear. Let the three decision criteria be aesthetics, comfort, and cost as depicted in Figure 2.3. Aesthetics is a subjective criterion, whereas clothing comfort can have subjective evaluation through user trial or objective measurement by

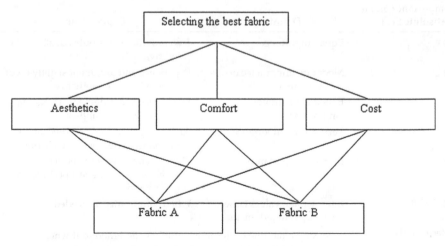

FIGURE 2.3
Decision problem for selecting the best fabric.

TABLE 2.2

Pair-Wise Comparison Matrix of Criteria

	Aesthetics	Comfort	Cost
Aesthetics	1	3	5
Comfort	1/3	1	3
Cost	1/5	1/3	1
Sum	1.533	4.333	9

evaluating air permeability, moisture vapor transmission rate, and thermal resistance. As comfort itself has multiple facets, it can further be divided into related subcriteria like air permeability, moisture vapor transmission, thermal resistance, etc. To make the situation simpler, only air permeability is considered here as a measure of comfort. The pair-wise comparisons of three criteria are shown in Table 2.2. The decision maker perceives that aesthetics is moderately dominant over comfort (score of 3), and comfort is moderately dominant over cost (score of 3). Aesthetics is strongly dominant over cost (score of 5). The matrix after normalization with the sum of columns is presented in Table 2.3.

Alternatively, by using Equation 2.1, the geometric mean of elements in the rows of the pair-wise comparison matrix, shown in Table 2.3, is found to be as follows:

$$\text{Aesthetics} = (1 \times 3 \times 5)^{1/3} = 2.466$$

$$\text{Comfort} = \left(\frac{1}{3} \times 1 \times 3\right)^{1/3} = 1$$

$$\text{Cost} = \left(\frac{1}{5} \times \frac{1}{3} \times 1\right)^{1/3} = 0.406$$

Normalizing the geometric means of three rows yields the criteria weights. So, the weight of aesthetics, comfort, and cost are 2.466/3.872, 1/3.872 and

TABLE 2.3

Normalization of Pair-Wise Comparison Matrix

	Aesthetics	Comfort	Cost	Row Average (Criteria Weight)
Aesthetics	0.652	0.692	0.556	0.634
Comfort	0.217	0.231	0.333	0.260
Cost	0.131	0.077	0.111	0.106
Sum	1	1	1	1

0.405/3.872—that is, 0.637, 0.258, and 0.105. These weights are practically the same as those obtained earlier and presented in Table 2.3.

For checking the consistency of judgement, matrices C_3 and C_4 are calculated such that $C_3 = C_1 \times C_2$ and $C_4 = (C_3/C_2)$, where $C_2 = [W_1, W_2, \ldots, W_N]^T$ and $W_1, W_2, \ldots, W_N$ are the weights of criterion 1, 2, $\ldots$, N, respectively. The principal eigenvector (λ_{max}) of the original pair-wise comparison matrix (C_1) is calculated from the average of matrix C_4. To check the consistency in pair-wise comparison, consistency index (CI) and consistency ratio (CR) are calculated using the following equation:

$$CI = \frac{\lambda_{max} - N}{N - 1} \quad \text{and} \quad CR = \frac{CI}{RCI} \tag{2.2}$$

where RCI is the random consistency index, and its value is shown in Table 2.4. If the value of CR is 0.1 or less, then the judgment is considered to be consistent and acceptable. Otherwise the decision maker has to make some changes in the entry of the pair-wise comparison matrix.

Similarly, N numbers of pair-wise comparison matrices, one for each criterion, of $M \times M$ order are formed where each alternative is compared with its competitors and pair-wise comparison is made with respect to each of the decision criterion. The eigenvectors of each of these "N" matrices represent the performance scores of alternatives in the corresponding criteria and form a column of the final decision matrix. The decision matrix is shown in Table 2.5.

In the case of the original AHP proposed by Saaty (1980, 1983, 1986), the sum of each column in the decision matrix is 1. This implies that $\sum_{i=1}^{M} a_{ij} = 1$.

TABLE 2.4

Random Consistency Index (RCI) Values for Different Numbers of Elements

N	1	2	3	4	5	6	7	8	9
RCI	0	0	0.58	0.90	1.12	1.24	1.32	1.41	1.45

TABLE 2.5

Decision Matrix of AHP

	Criteria				
	C_1	C_2	C_3	...	C_N
Alternatives	(W_1)	(W_2)	(W_3)	...	(W_N)
A_1	a_{11}	a_{12}	a_{13}	...	a_{1N}
A_2	a_{21}	a_{22}	a_{23}	...	a_{2N}
A_3	a_{31}	a_{32}	a_{33}	...	a_{3N}
...	...	...	...	...	...
A_M	a_{M1}	a_{M2}	a_{M3}	...	a_{MN}

Step 4

In this step, the final priority of alternatives is determined by considering the alternative scores (a_{ij}) in each criterion and the weight of the corresponding criterion (W_j) using the following equation:

$$A_{AHP} = \max \sum_{j=1}^{N} a_{ij}.W_j \quad \text{for } i = 1, 2, 3, \ldots, M \tag{2.3}$$

Continuing with the example of best fabric selection, fabrics A and B will be compared with each other with respect to each of the three decision criteria—that is, aesthetics, comfort, and cost. The pair-wise comparison matrix of two fabrics with respect to aesthetics is shown in Table 2.6. The priority score of fabrics with respect to aesthetics is shown in Table 2.7.

As the two other decision criteria are objective in nature, pair-wise comparison can be done away with, and the scores of fabrics A and B with respect to decision criteria can directly be entered in the final decision matrix. Let it be assumed that air permeability of fabrics A and B is 100 and 200 cm³/ cm²/s, respectively, and their costs are 15 and 10 $/m², respectively. As in the case of air permeability and cost, where higher and lower values, respectively, are desirable, the normalization should be done as follows:

Normalized value$_i$

$$= \frac{\text{Score of the } i\text{th alternative}}{\text{Maximum score among the alternatives}} \text{ (for all benefit criteria)}$$

Normalized value$_i$

$$= \frac{\text{Minimum score among the alternatives}}{\text{Score of the } i\text{th alternative}} \text{ (for all cost criteria)}$$

TABLE 2.6

Pair-Wise Comparison Matrix of Alternatives (Fabrics) with Respect to Criterion (Aesthetics)

	Fabric A	Fabric B
Fabric A	1	2
Fabric B	1/2	1

TABLE 2.7

Priority Score of Fabrics with Respect to Aesthetics

	Fabric A	Fabric B	Priority Score
Fabric A	0.667	0.667	0.667
Fabric B	0.333	0.333	0.333

TABLE 2.8

Decision Matrix for Fabric Selection

	Aesthetics (0.634)	Comfort (0.260)	Cost (0.106)	Overall Priority Score
Fabric A	0.667	100/200	10/15	0.624
Fabric B	0.333	200/200	10/10	0.577

The final decision matrix is shown in Table 2.8.

So, fabric A is preferred over fabric B as it has a higher overall priority score.

2.2.1 Importance of Hierarchical Structure

Let us assume that a MCDM problem has five criteria and each of them has three subcriteria making 15 subcriteria in total. So, in a hierarchically structured problem, having one level for criteria and the next level for subcriteria, the total number of pair-wise comparison judgments will be as follows:

$$^5C_2 + 5 \times {}^3C_2 = \frac{5 \times (5-1)}{2} + 5 \times \frac{3 \times (3-1)}{2} = 10 + 15 = 25$$

So, the decision maker will have to make 10 and 15 judgements at the level of criteria and subcriteria, respectively. However, if the level of criteria is removed and 15 subcriteria are compared, in a pair-wise manner, with respect to the goal of the problem, then the decision maker will have to make the following judgments:

$$^{15}C_2 = \frac{15 \times (15-1)}{2} = 105$$

Therefore, hierarchical structure reduces the number of pair-wise judgments and thus the possibility of inconsistent judgment is reduced.

Now, let us take the example depicted in Figure 2.2. Let the weight of criterion 1 with respect to the goal (global weight) be 0.35 and the weight of subcriteria 1A and 1B with respect to criterion 1 (local weight) be 0.6 and 0.4, respectively. Therefore, the global weight of subcriteria 1A and 1B with respect to the goal of the problem will be 0.35 × 0.6 and 0.35 × 0.4—that is, 0.21 and 0.14, respectively.

2.2.2 Rank Reversal in Analytic Hierarchy Process

The most common criticism of MCDM methods, including AHP, is their susceptibility to rank reversal (Belton and Gear, 1983; Saaty and Vargas, 1984; Triantaphyllou and Mann, 1994; Triantaphyllou, 2001). There can be various types of rank reversals as follows:

- *Type 1:* The relative ranking of alternatives changes when one identical or near-identical copy of a nonoptimal alternative is added in the fray.

- *Type 2*: When a bigger problem is decomposed into multiple smaller problems, keeping the criteria weights the same, the ranking of a smaller problem may not be in agreement with that of a bigger problem.
- *Type 3*: When a bigger problem is decomposed to multiple smaller problems, keeping the criteria weights the same, the ranking of a smaller problem may not show the transitivity property. For example, if only two alternatives are considered at a time, it may yield $A > B$ and $B > C$. However, when only A and C are considered, keeping the same criteria and weights, it may lead to $C > A$.

Example of Type 1 Rank Reversal

Belton and Gear (1983) pointed out that reversal of a relative ranking of alternatives may occur in AHP when a new alternative is added or deleted. They opined that the criteria weights may change due to the introduction of a new alternative, and then the rank reversal becomes obvious. However, even without any change in the criteria weights, AHP can show rank reversals. They presented an example having three decision criteria and three alternatives. All of the criteria were equally important, and therefore, the weight of each criterion was 1/3. The pair-wise comparison matrices of three alternatives, for three decision criteria, were as follows:

	Criterion 1			Criterion 2			Criterion 3		
Alternative A	1	1/9	1	1	9	9	1	8/9	8
Alternative B	9	1	9	1/9	1	1	9/8	1	9
Alternative C	1	1/9	1	1/9	1	1	1/8	1/9	1

The overall priority for alternatives A, B, and C are 0.45, 0.47, 0.08, respectively. Therefore, $B > A$ or B is preferred over A. Now, a new alternative D, which is identical to B, is introduced in fray. The pair-wise comparison matrices now become as shown:

	Criterion 1				Criterion 2				Criterion 3			
Alternative A	1	1/9	1	1/9	1	9	9	9	1	8/9	8	8/9
Alternative B	9	1	9	1	1/9	1	1	1	9/8	1	9	1
Alternative C	1	1/9	1	1/9	1/9	1	1	1	1/8	1	1	1/9
Alternative D	9	1	9	1	1/9	1	1	1	9	1	9	1

Now, the overall priority for four alternatives A, B, C, and D are 0.37, 0.29, 0.06, and 0.29, respectively. So, $A > B$ or A is preferred over B, which implies rank reversal. Belton and Gear (1983) suggested that change in local priorities, with respect to criteria, with the introduction of new alternatives is the root cause behind the rank reversal in AHP. They suggested that the method of normalization, which makes the sum of each column as 1, is not robust in the case of AHP. They proposed a new method of normalization of alternative scores by dividing each alternative score with the maximum value in the corresponding vector. This variant of AHP is known as revised AHP.

Saaty and Vargas (1984) countered the method of normalization proposed by Belton and Gear (1983) with a new example where even the normalization with the maximum value of the weight vector led to rank reversal after introduction of a new alternative. They also opined that rank reversal can also be seen as a good thing as the introduction of a new alternative changes the old preference order. According to them, the rank preservation after the introduction of new alternatives should not be treated as a dogma. They further argued that truth is a human construct, and it may change with the availability of new information. The relationship between the two alternatives is only tentative, and it may change with the arrival of a new alternative. For example, a decision maker is presented with two cars X and Y. While car Y is costlier than X, the former offers better comfort than the latter. The cost-conscious decision maker might prefer car X over car Y. Now, a new car Z is presented to the decision maker. Car Z is three times costlier than car Y; however, it offers only marginally better comfort than car Y. Now, car Y may be the most preferred one as the decision maker may think that by paying slightly extra he or she is getting almost the best possible comfort. Saaty and Vargas (1984) made the following observations regarding rank reversals:

- If the new alternative is dominated by the least preferred alternative in all criteria, rank reversal is unlikely to happen.
- If the new alternative dominates the most preferred alternative in all criteria, rank reversal is unlikely to happen.
- If the new alternative lies between two specific alternatives in all criteria, then the new alternative will rank in between the two specific alternatives. However, rank reversal may occur elsewhere.

Some researchers (Schenkerman, 1994; Wang and Elhag, 2006) have proposed approaches to avoid rank reversal in AHP by preserving the local priorities of alternatives when a new alternative is introduced.

Example of Type 2 Rank Reversal

Triantaphyllou (2001) showed that both the original AHP and revised AHP are susceptible to rank reversal. The example used for original AHP is shown in Table 2.9.

When three alternatives are considerd together, $C > B > A$. Now, the problem is decomposed considering only two alternatives B and C as shown in Table 2.10. From the priority score, $B > C$, which is not in agreement with the ranking obtained earlier with all three alternatives—that is, $C > B > A$.

Triantaphyllou (2001) also showed that revised AHP is also susceptible to this kind of rank reversal. The example used for revised AHP is presented in Table 2.11. Based on the priority score, $B > C > A$. Now, if the problem is decomposed and only two alternatives B and C are considered as shown in Table 2.12, based on the priority score, $C > B$, which is not in agreement with the ranking obtained earlier with all three alternatives—that is, $B > C > A$.

TABLE 2.9

Example of Rank Reversal in Original AHP

Alternatives	Criterion 1 Weight (2/7)	Criterion 2 Weight (2/7)	Criterion 3 Weight (3/7)	Priority Score
A	9/19	2/12	2/7	0.305
B	5/19	1/12	4/7	0.344
C	5/19	9/12	1/7	0.351

Source: Triantaphyllou, E. 2001. *Journal of Multi-Criteria Decision Analysis*, 10, 11–25.

TABLE 2.10

Rank Reversal in Original AHP after Decomposition

Alternatives	Criterion 1 Weight (2/7)	Criterion 2 Weight (2/7)	Criterion 3 Weight (3/7)	Priority Score
B	5/10	1/10	4/5	0.514
C	5/10	9/10	1/5	0.486

Source: Triantaphyllou, E. 2001. *Journal of Multi-Criteria Decision Analysis*, 10, 11–25.

TABLE 2.11

Example of Rank Reversal in Revised AHP

Alternatives	Criterion 1 Weight (4/22)	Criterion 2 Weight (9/22)	Criterion 3 Weight (9/22)	Priority Score
A	9/9	5/8	2/8	0.540
B	1/9	8/8	5/8	0.685
C	8/9	2/8	8/8	0.673

Source: Triantaphyllou, E. 2001. *Journal of Multi-Criteria Decision Analysis*, 10, 11–25.

TABLE 2.12

Rank Reversal in Revised AHP after Decomposition

Alternatives	Criterion 1 Weight (4/22)	Criterion 2 Weight (9/22)	Criterion 3 Weight (9/22)	Priority Score
B	1/8	8/8	5/8	0.687
C	8/8	2/8	8/8	0.693

Source: Triantaphyllou, E. 2001. *Journal of Multi-Criteria Decision Analysis*, 10, 11–25.

2.2.3 Multiplicative Analytic Hierarchy Process

Triantaphyllou and Mann (1994) demonstrated that AHP and revised AHP might yield different rankings of alternatives than the ranking that would result if the actual relative importance of criteria were known. The probability of this incorrect ranking increases drastically as the number of alternatives increases. Moreover, as mentioned earlier, additive variants of AHP do not always exhibit the transitivity property. For example, if $A_1 > A_2$ and $A_2 > A_3$, then AHP and revised AHP fail to guarantee that $A_1 > A_3$.

Lootsma (1993) proposed the multiplicative version of AHP, which is similar to the WPM. In multiplicative AHP (MAHP), the relative priority or P_i of alternative A_i is derived by using the following formula:

$$P_i = \prod_{j=1}^{N} (a_{ij})^{W_j}, \quad \text{for } i = 1, 2, 3, \dots, M \tag{2.4}$$

The relative priority derived by using the previous formula is independent of the way in which relative performance values a_{ij} have been normalized. Triantaphyllou (2001) demonstrated that the ranking produced by MAHP is quite robust and free from some of the ranking inconsistencies.

2.3 Fuzzy Analytic Hierarchy Process

One of the drawbacks of AHP is that it fails to translate the imprecision and uncertainty in the pair-wise comparison matrix as it deals only with crisp numbers (one through nine and their reciprocals). This can be overcome by the integration of fuzzy logic with AHP. The fuzzy analytic hierarchy process (FAHP) works in a similar fashion as crisp AHP. However, the basic difference lies in the comparison of elements (criteria or alternatives) and assigning their relative importance in relation to the element at the immediate upper level. Instead of using a crisp number, triangular fuzzy numbers are used in most of the cases to represent the vagueness in the perception of the decision maker.

2.3.1 Fuzzy Numbers and Their Operations

A fuzzy number can be represented in many ways (triangular, trapezoidal, Gaussian, etc.). The use of a triangular fuzzy number (TFN) is the simplest way to represent imprecision. A fuzzy number M on R is to be a TFN if its membership function $\mu_M(x):R \rightarrow [0, 1]$ is equal to the following:

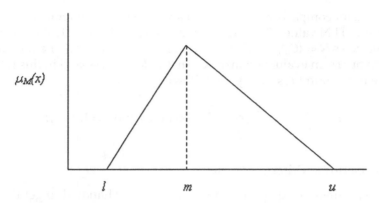

FIGURE 2.4
Triangular fuzzy number.

$$\mu_M(x) = \begin{cases} \dfrac{x-l}{m-l}, & x \in [l,m], \\ \dfrac{x-u}{m-u}, & x \in [m,u], \\ 0, & \text{otherwise} \end{cases} \tag{2.5}$$

where $l \le m \le u$, l and u stand for the lower and upper values, and m is the modal value. A TFN can also be depicted as shown in Figure 2.4.

The basic operations of TFN are as follows:

$$\left. \begin{aligned} \bar{n}_1 \oplus \bar{n}_2 &= (n_{1l} + n_{2l}, n_{1m} + n_{2m}, n_{1u} + n_{2u}), && \text{for addition} \\ \bar{n}_1 \ominus \bar{n}_2 &= (n_{1l} - n_{2l}, n_{1m} - n_{2m}, n_{1u} - n_{2u}), && \text{for subtraction} \\ \bar{n}_1 \otimes \bar{n}_2 &= (n_{1l} \times n_{2l}, n_{1m} \times n_{2m}, n_{1u} \times n_{2u}), && \text{for multiplication} \\ \frac{\bar{n}_1}{\bar{n}_2} &= \left(\frac{n_{1l}}{n_{2u}}, \frac{n_{1m}}{n_{2m}}, \frac{n_{1u}}{n_{2l}} \right), && \text{for division} \\ \frac{1}{\bar{n}_1} &= \left(\frac{1}{n_{1u}}, \frac{1}{n_{1m}}, \frac{1}{n_{1l}} \right), && \text{for inverse} \end{aligned} \right\} \tag{2.6}$$

where $\bar{n}_1 = (n_{1l}, n_{1m}, n_{1u})$ and $\bar{n}_2 = (n_{2l}, n_{2m}, n_{2u})$ represent two TFNs with lower, modal, and upper values.

2.3.2 Developing Decision Hierarchy and Constructing Fuzzy Comparison Matrix

The decision hierarchy of the problem is developed in accordance to the study conducted. The expert's opinion and judgment in linguistic terms are considered for comparison of all criteria and subcriteria and transformed into the definite TFN values. Therefore, a fuzzy comparison evaluation matrix is constructed as $N = (C_{ij})_{n \times m}$, where C_{ij} represents the entries in the constructed fuzzy comparison evaluation matrix—that is, (a_{ij}, b_{ij}, c_{ij})—and in this relation, positive fuzzy numbers satisfy the following property:

$$a_{ij} = \frac{1}{a_{ji}}, b_{ij} = \frac{1}{b_{ji}}, c_{ij} = \frac{1}{c_{ji}}, \quad \text{where } i \text{ and } j = 1, 2, \dots, z$$

2.3.3 Computing Criteria Weights

The fuzzy pair-wise comparison matrix is further handled to get the crisp values of criteria weights. For this purpose, the fuzzy extent analysis method proposed by Chang (1996) is popularly used.

Let $X = (x_1, x_2, x_3, \dots, x_n)$ be an object set, and $U = (u_1, u_2, u_3, \dots, u_n)$ be a goal set. Each object is considered, and extent analysis for each goal g_i is carried out. Therefore, m extent analysis values for each object can be obtained as follows:

$$M_{gi}^1, M_{gi}^2, M_{gi}^3, \dots, M_{gi}^m, i = 1, 2, \dots, n$$

where all are TFNs:

$$M_{gi}^j (j = 1, 2, 3, \dots, m)$$

The value of fuzzy synthetic extent with respect to the ith object is defined as follows:

$$S_i = \sum_{j=1}^m M_{gi}^j \otimes \left[\sum_{i=1}^n \sum_{j=1}^m M_{gi}^j \right]^{-1} \tag{2.7}$$

To obtain $\sum_{j=1}^m M_{gi}^j$, it is necessary to perform the fuzzy addition operation of m extent analysis values for a particular matrix as follows:

$$\sum_{j=1}^m M_{gi}^j = \left(\sum_{j=1}^m l_j, \sum_{j=1}^m m_j, \sum_{j=1}^m u_j \right) \tag{2.8}$$

To obtain $\left[\sum_{i=1}^n \sum_{j=1}^m M_{gi}^j \right]^{-1}$, fuzzy addition should be performed on M_{gi}^j as follows:

$$\sum_{i=1}^n \sum_{j=1}^m M_{gi}^j = \left(\sum_{i=1}^n l_i, \sum_{i=1}^n m_i, \sum_{j=1}^n u_i \right) \tag{2.9}$$

The inverse of the vector shown is calculated such that

$$\left[\sum_{i=1}^{n}\sum_{j=1}^{m}M_{gi}^{j}\right]^{-1} = \left(\frac{1}{\sum_{i=1}^{n}u_i}, \frac{1}{\sum_{i=1}^{n}m_i}, \frac{1}{\sum_{i=1}^{n}l_i}\right) \tag{2.10}$$

The degree of possibility of two fuzzy numbers $M_2 = (l_2, m_2, u_2) \geq M_1(l_1, m_1, u_1)$ is defined as follows:

$$V(M_2 \geq M_1) = hgt(M_1 \cap M_2) = \mu_{M_2}(d)$$

$$= \begin{cases} 1, & \text{if } m_2 \geq m_1 \\ 0, & \text{if } l_1 \geq u_2 \\ \dfrac{l_1 - u_2}{(m_2 - u_2) - (m_1 - l_1)}, & \text{otherwise} \end{cases} \tag{2.11}$$

where d is the ordinate of the highest intersection point between μ_{M_1} and μ_{M_2} as depicted in Figure 2.5.

For the comparison between two fuzzy numbers M_1 and M_2, both values indicating $V(M_1 \geq M_2)$ and $V(M_2 \geq M_1)$ are needed. The degree of possibility for a convex fuzzy number to be greater than k convex fuzzy numbers $M_i(i = 1, 2, \ldots, k)$ can be defined as

$$\begin{aligned} V(M \geq M_1, M_2, \ldots, M_k) \\ = V[(M \geq M_1) \text{ and } (M \geq M_2) \text{ and } \ldots \text{ and } (M \geq M_k)] \\ = \min V(M \geq M_i), i = 1, 2, \ldots, k \end{aligned} \tag{2.12}$$

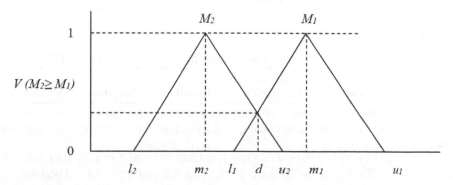

FIGURE 2.5
Intersection between two fuzzy numbers M_1 and M_2.

If $d'(A_i) = \min V(S_i \geq S_k)$ for $k = 1, 2, \ldots, n$, then the weight vector is given as follows:

$$W' = (d'(A_1), d'(A_2), \ldots, d'(A_n))^T \tag{2.13}$$

where $A_i(i = 1, 2, \ldots, n)$ are n elements.

After normalization, normalized weight vectors are obtained as follows:

$$W = (d(A_1), d(A_2), \ldots, d(A_n))^T \tag{2.14}$$

where W is a nonfuzzy or crisp number.

2.3.4 Example of Fuzzy Analytic Hierarchy Process Application

Let a case of cotton fiber selection be considered under fuzzy environment. The decision maker has some vague or imprecise idea about the relative importance of five cotton fiber properties (length, length uniformity, strength, micronaire, and elongation) in determining the ring spun yarn tenacity. Table 2.13 shows the fuzzy numbers used to represent linguistic expressions while comparing different elements of the pair-wise comparison matrix (Wang et al., 2007). Table 2.14 presents the pair-wise comparison matrix of cotton fiber properties using fuzzy numbers.

TABLE 2.13

Fuzzy Numbers Equivalent to Linguistic Expressions

Judgment	Fuzzy Number
Approximately equal	$1/2, 1, 2$
Approximately x times more significant	$x - 1, x, x + 1$
Approximately x times less significant	$1/(x + 1), 1/x, 1/(x - 1)$
Between y and z times more significant	$y, (y + z)/2, z$
Between y and z times less significant	$1/z, 2/(y + z), 1/y$

TABLE 2.14

Pair-Wise Comparison Matrix of Cotton Fiber Properties Using Fuzzy AHP

Fiber Parameters	Length	Length Uniformity	Strength	Elongation	Micronaire
Length	1.00, 1.00, 1.00	1.00, 2.00, 3.00	1.00, 2.00, 3.00	2.00, 3.00, 4.00	2.00, 3.00, 4.00
Length uniformity	0.33, 0.50, 1.00	1.00, 1.00, 1.00	0.33, 0.50, 1.00	1.00, 2.00, 3.00	1.00, 2.00, 3.00
Strength	0.33, 0.50, 1.00	1.00, 2.00, 3.00	1.00, 1.00, 1.00	2.00, 3.00, 4.00	2.00, 3.00, 4.00
Micronaire	0.25, 0.33, 0.50	0.33, 0.50, 1.00	0.25, 0.33, 0.50	1.00, 2.00, 3.00	1.00, 1.00, 1.00
Elongation	0.25, 0.33, 0.50	0.33, 0.50, 1.00	0.25, 0.33, 0.50	1.00, 1.00, 1.00	0.33, 0.50, 1.00

Source: Wang, L. et al. 2007. *International Journal of Production Economics*, 107, 151–163.

The sum of entries of rows is as follows:

Length: (7.00, 11.00, 15.00)
Length uniformity: (3.67, 6.00, 9.00)
Strength: (6.33, 0.33, 13.00)
Micronaire: (2.83, 4.17, 6.00)
Elongation: (2.17, 2.67, 4.00)

The cumulative sum of these five fuzzy numbers is (22, 33.33, 47.00). Now, following the steps mentioned in Equations 2.7 to 2.10, the fuzzy weights of fiber parameters are obtained:

$$\text{Length} = (7.00, 11.00, 15.00) \times \left(\frac{1}{47.0}, \frac{1}{33.33}, \frac{1}{22.00} \right)$$
$$= (0.15, 0.33, 0.68)$$

$$\text{Strength} = (6.33, 0.33, 13.00) \times \left(\frac{1}{47.0}, \frac{1}{33.33}, \frac{1}{22.00} \right)$$
$$= (0.13, 0.28, 0.59)$$

$$\text{Length uniformity} = (3.67, 6.00, 9.00) \times \left(\frac{1}{47.0}, \frac{1}{33.33}, \frac{1}{22.00} \right)$$
$$= (0.08, 0.18, 0.41)$$

$$\text{Micronaire} = (2.83, 4.17, 6.00) \times \left(\frac{1}{47.0}, \frac{1}{33.33}, \frac{1}{22.00} \right)$$
$$= (0.06, 0.13, 0.27)$$

$$\text{Elongation} = (2.17, 2.67, 4.00) \times \left(\frac{1}{47.0}, \frac{1}{33.33}, \frac{1}{22.00} \right)$$
$$= (0.05, 0.08, 0.18)$$

Following the steps explained in Equations 2.13 and 2.14, normalized crisp weights of cotton fiber properties are obtained, which are presented in Table 2.15. As the cotton fiber properties are measured objectively by high-volume instruments (HVIs), no pair-wise comparison is required between various types of cotton alternatives with respect to cotton fiber properties. The measured values of various cotton fiber alternatives can directly be entered into the decision matrix, and the final priority of different alternatives can be determined.

TABLE 2.15

Weights of Cotton Fiber Properties after Pairwise Comparison Evaluation

Fiber Parameters	Fuzzy Weight	Crisp Weight	Normalized Crisp Weight	Rank
Length	0.15, 0.33, 0.68	1	0.314	1
Length uniformity	0.08, 0.18, 0.41	0.638	0.200	2
Strength	0.13, 0.29, 0.59	0.909	0.285	3
Elongation	0.06, 0.13, 0.27	0.382	0.120	4
Micronaire	0.05, 0.08, 0.18	0.256	0.080	5

2.4 Summary

AHP is one of the most popular multicriteria decision-making methods. It has been applied extensively in every possible facet of decision-making in engineering, social sciences, and management. In the domain of textiles, AHP has been used for the ranking of cotton fibers and selection of fabrics. The fuzzy AHP can be used in situations where the decision maker has imprecise or vague knowledge about the decision criteria or alternatives. Though additive variants of AHP are susceptible to rank reversals, they are extremely useful for the selection of the best alternative or ranking of alternatives based on multiple decision criteria.

References

Belton, V. and Gear, T. 1983. On a shortcoming of Saaty's method of analytic hierarchies. *Omega*, 11, 228–230.

Chang, D. Y. 1996. Application of the extent analysis method on fuzzy AHP. *European Journal of Operations Research*, 95, 649–655.

Dyer, J. S. 1990a. A clarification of "remarks on the analytic hierarchy process." *Management Science*, 36, 274–275.

Dyer, J. S. 1990b. Remarks on the analytic hierarchy process. *Management Science*, 36, 249–258.

Harker, P. T. and Vargas, L. G. 1990. Reply to "remarks on the analytic hierarchy process." *Management Science*, 36, 269–273.

Lootsma, F. A. 1993. Scale sensitivity in the multiplicative AHP and SMART. *Journal of Multi-Criteria Decision Analysis*, 2, 87–110.

Saaty, T. L. 1980. *The Analytic Hierarchy Process*. McGraw-Hill, New York.

Saaty, T. L. 1983. Priority setting in complex problems. *IEEE Transactions on Engineering Management*, 30, 140–155.

Saaty, T. L. 1986. Axiomatic foundation of analytic hierarchy process. *Management Science*, 32, 841–855.

Saaty, T. L. and Vargas, L. G. 1984. The legitimacy of rank reversal. *Omega*, 12, 513–516.

Schenkerman, S. 1994. Avoiding rank reversals in AHP decision support models. *European Journal of Operations Research*, 74, 407–419.

Triantaphyllou, E. 2001. Two new cases of rank reversals when the AHP and some of its additive variants are used that do not occur with multiplicative AHP. *Journal of Multi-Criteria Decision Analysis*, 10, 11–25.

Triantaphyllou, E. and Mann, S. H. 1994. A computational evaluation of the original and revised analytic hierarchy process. *Computers and Industrial Engineering*, 26, 609–618.

Wang, L., Chu, J. and Wu, J. 2007. Selection of optimum maintenance strategies based on a fuzzy analytic hierarchy process. *International Journal of Production Economics*, 107, 151–163.

Wang, Y. M. and Elhag, T. M. S. 2006. An approach to avoiding rank reversal in AHP. *Decision Support Systems*, 42, 1474–1480.

3

Technique for Order of Preference by Similarity to Ideal Solution (TOPSIS)

3.1 Introduction

Multicriteria decision-making (MCDM) is the method of finding the best option or ranking the entire set of options from all of the feasible alternatives in the presence of multiple, usually conflicting criteria. A MCDM problem is usually represented in the form of a matrix known as the decision matrix, as shown in Table 3.1. A decision matrix is a $(m \times n)$ matrix in which element x_{ij} denotes the performance of alternative A_i when it is evaluated in terms of decision criterion C_j, where $i = 1,2,3, \dots ,m$ and $j = 1,2,3, \dots ,n$. Numerical weight (w_j) is attached to each criterion based on its relative importance such that $\sum_{j=1}^{n} w_j = 1$. All of the elements in the decision matrix must be normalized to eliminate the units of different criteria.

The technique for order preference by similarity to ideal solution (TOPSIS) is one of the widely used methods of MCDM. The fundamental idea of this method is that the chosen alternative should have the shortest distance from the ideal solution and the farthest distance from the negative-ideal solution.

In many real-life situations, due to the vague concepts frequently represented in decision data, crisp values are inadequate to represent a decision matrix of a MCDM problem. Considering the vagueness in the decision data, linguistic variables are used to form a fuzzy decision matrix. A fuzzy-TOPSIS method is an extension of the concept of TOPSIS in the fuzzy environment.

3.2 TOPSIS Methodology

TOPSIS was developed by Hwang and Yoon (1981). This method is based on the assumption that each criterion has a tendency to monotonically increase or decrease utility; hence, it is easy to define ideal and negative-ideal solutions. The ideal and negative-ideal solutions are hypothetically the best and worst solutions, respectively. The TOPSIS leads to a solution that is not only closest

TABLE 3.1

Decision Matrix for a MCDM Problem

Criteria		C_1	C_2	..	C_j	..	C_n
Weights		w_1	w_2	..	w_j	..	w_n
	A_1	x_{11}	x_{12}	..	x_{1j}	..	x_{1n}
	A_2	x_{21}	x_{22}	..	x_{2j}	..	x_{2n}
	$\vdots$	$\vdots$	$\vdots$	..	$\vdots$	..	$\vdots$
Alternatives	A_i	x_{i1}	x_{i2}	..	x_{ij}	..	x_{in}
	$\vdots$	$\vdots$	$\vdots$	..	$\vdots$	..	$\vdots$
	A_m	x_{m1}	x_{m2}	..	$\vdots$	..	x_{mn}

to the hypothetical best, but also it is farthest from the hypothetical worst. The principles of this method are discussed in the following steps.

Step 1: Construct the Decision Matrix

This step produces decision matrix D of criteria and alternatives based on the information available regarding the MCDM problem. If the number of alternatives is m, and the number of criteria is n, then a decision matrix having an order of $m \times n$ can be represented as follows:

$$D = \begin{bmatrix} x_{11} & x_{12} & \cdots & x_{1n} \\ x_{21} & x_{22} & \cdots & x_{2n} \\ \cdots & \cdots & \cdots & \cdots \\ x_{m1} & x_{m2} & \cdots & x_{mn} \end{bmatrix} \quad (3.1)$$

where x_{ij} denotes the performance measure of the ith alternative in terms of the jth criterion.

Step 2: Construct the Normalized Decision Matrix

In this step, the decision matrix is converted to a normalized decision matrix R. An element r_{ij} of the normalized decision matrix is estimated as follows:

$$r_{ij} = \frac{x_{ij}}{\sqrt{\sum_{k=1}^{m} x_{kj}^2}} \quad (3.2)$$

and

$$R = \begin{bmatrix} r_{11} & r_{12} & \cdots & r_{1n} \\ r_{21} & r_{22} & \cdots & r_{2n} \\ \cdots & \cdots & \cdots & \cdots \\ r_{m1} & r_{m2} & \cdots & r_{mn} \end{bmatrix} \quad (3.3)$$

Step 3: Construct the Weighted Normalized Matrix

The weighted normalized matrix is obtained by multiplying each column of the normalized decision matrix R with the associated criteria weight corresponding to that column. Hence, a weighted normalized matrix V is represented as follows:

$$V = RW \tag{3.4}$$

where

$$
W = \begin{bmatrix}
w_1 & 0 & 0 & 0 & \cdots & 0 \\
0 & w_2 & 0 & 0 & \cdots & 0 \\
\cdots & \cdots & \cdots & \cdots & \cdots & \cdots \\
0 & 0 & 0 & \cdots & \cdots & w_n
\end{bmatrix} \tag{3.5}
$$

and $\sum w_i = 1$.

Thus, the expression of V becomes

$$
V = \begin{bmatrix}
w_1 r_{11} & w_2 r_{12} & w_3 r_{13} & \cdots & w_n r_{1n} \\
w_1 r_{21} & w_2 r_{22} & w_3 r_{23} & \cdots & w_n r_{2n} \\
\cdot & \cdot & \cdot & \cdots & \cdot \\
\cdot & \cdot & \cdot & \cdots & \cdot \\
\cdot & \cdot & \cdot & \cdots & \cdot \\
w_1 r_{m1} & w_2 r_{m2} & w_3 r_{m3} & \cdots & w_n r_{mn}
\end{bmatrix} \tag{3.6}
$$

An element v_{ij} of the weighted normalized decision matrix is estimated as follows:

$$v_{ij} = w_j . r_{ij} \tag{3.7}$$

Step 4: Determine the Ideal and the Negative-Ideal Solutions

The ideal solution, denoted by A^*, is defined as follows:

$$
\begin{aligned}
A^* &= \left\{ \left(\max_i v_{ij} \mid j \in J \right), \left(\min_i v_{ij} \mid j \in J' \right), i = 1, 2, 3, \ldots, m \right\} \\
&= \left\{ v_1^*, v_2^*, \ldots, v_n^* \right\}
\end{aligned} \tag{3.8}
$$

The negative-ideal solution, denoted as A^-, is defined by

$$
\begin{aligned}
A^- &= \left\{ \left(\min_i v_{ij} \mid j \in J \right), \left(\max_i v_{ij} \mid j \in J' \right), i = 1, 2, 3, \ldots, m \right\} \\
&= \left\{ v_1^-, v_2^-, \ldots, v_n^- \right\}
\end{aligned} \tag{3.9}
$$

where $J = \{j = 1,2,3, \dots, n$ and j is associated with benefit criteria$\}$ and $J' = \{j = 1,2,3, \dots, n$ and j is associated with cost criteria$\}$.

For the benefit criteria, the decision maker wants to have the maximum value among the alternatives. On the contrary, in the case of cost criteria, the decision maker wants to have the minimum value among the alternatives. Hence, A^* points to the best possible alternative or ideal solution. Similarly, A^- specifies the worst possible alternative, or negative ideal solution.

Step 5: Calculate the Separation Measure

In this step, the n-dimensional Euclidean distance is used to measure the separation distances of each alternative from the ideal solution as well as the negative ideal solution. Thus, the separation distance of the ith alternative from the ideal solution, which is denoted by S_{i^*}, is expressed as follows:

$$S_{i^*} = \sqrt{\sum_{j=1}^{n}\left(v_{ij} - v_{j^*}\right)^2}, \quad \text{for } i = 1,2,3,\dots,m \qquad (3.10)$$

Similarly, the separation distance of the ith alternative from the negative-ideal solution, which is denoted as S_{i^-}, has the following expression:

$$S_{i^-} = \sqrt{\sum_{j=1}^{n}\left(v_{ij} - v_{j^-}\right)^2}, \quad \text{for } i = 1,2,3,\dots,m \qquad (3.11)$$

Figure 3.1 illustrates a typical example of the separation distances of the ith alternative from the ideal solution and the negative ideal solution.

Step 6: Calculate the Relative Closeness to the Ideal Solution

In this step, the relative closeness (C_{i^*}) value of an alternative A_i with respect to the ideal solution is calculated as follows:

$$C_{i^*} = \frac{S_{i^-}}{S_{i^*} + S_{i^-}} \qquad (3.12)$$

where $1 \geq C_{i^*} \geq 0$, and $i = 1,2,3,\dots m$. Typically, $C_{i^*} = 1$, if $A_i = A^*$, and $C_{i^*} = 0$, if $A_i = A^-$.

Step 7: Rank the Preference Order

All of the alternatives are now arranged in descending order according to the value of C_{i^*}. Therefore, the best alternative has the shortest distance from the ideal solution and the longest distance from the negative-ideal solution.

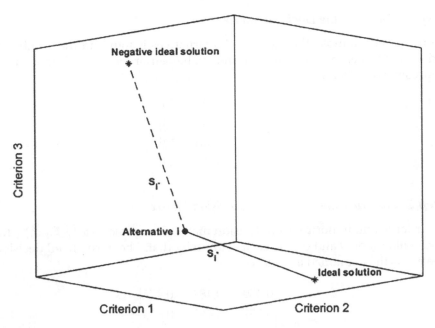

FIGURE 3.1
Separation measure from the ideal solution and the negative ideal solution.

3.3 Step-by-Step Working Principles of TOPSIS

Suppose that a MCDM problem involves four alternatives A_1, A_2, A_3, and A_4 and three criteria C_1, C_2, and C_3, where C_1 and C_2 are benefit criteria, and C_3 is the cost criterion. Let us also suppose that the relative weight of each criterion has equal importance. Given that the sum of relative weights is equal to one, the relative weight of each criterion is 1/3. The performance values of the four alternatives in terms of the three decision criteria are given in Table 3.2.

The following steps explain the working principles of the TOPSIS MCDM problem.

TABLE 3.2

Decision Matrix of the Problem

Criteria		C_1	C_2	C_3
Weights		1/3	1/3	1/3
	A_1	100	35	55
Alternatives	A_2	115	34	65
	A_3	90	40	52
	A_4	110	36	60

Step 1: Construct the Decision Matrix

In the given example, the number of alternatives is four and the number of criteria is three, thus the decision matrix has an order of 4×3 and can be represented as follows:

$$D = \begin{bmatrix} 100 & 35 & 55 \\ 115 & 34 & 65 \\ 90 & 40 & 52 \\ 110 & 36 & 60 \end{bmatrix}$$

Step 2: Construct the Normalized Decision Matrix

In order to obtain normalized decision matrix R, we calculate $(\sum_{k=1}^{m} x_{kj}^2)^{0.5}$ for each column of D and divide each column by that. The normalized decision matrix is thus obtained as follows:

$$R = \begin{bmatrix} 0.480 & 0.482 & 0.472 \\ 0.552 & 0.468 & 0.558 \\ 0.432 & 0.551 & 0.447 \\ 0.528 & 0.496 & 0.515 \end{bmatrix}$$

Step 3: Construct the Weighted Normalized Matrix

The relative weight matrix W can be expressed as

$$W = \begin{bmatrix} 1/3 & 0 & 0 \\ 0 & 1/3 & 0 \\ 0 & 0 & 1/3 \end{bmatrix}$$

We obtain the weighted normalized matrix V by multiplying R with W as follows:

$$V = \begin{bmatrix} 0.480 & 0.482 & 0.472 \\ 0.552 & 0.468 & 0.558 \\ 0.432 & 0.551 & 0.447 \\ 0.528 & 0.496 & 0.515 \end{bmatrix} \times \begin{bmatrix} 1/3 & 0 & 0 \\ 0 & 1/3 & 0 \\ 0 & 0 & 1/3 \end{bmatrix}$$

$$\therefore V = \begin{bmatrix} 0.160 & 0.161 & 0.157 \\ 0.184 & 0.156 & 0.186 \\ 0.144 & 0.184 & 0.149 \\ 0.176 & 0.165 & 0.172 \end{bmatrix}$$

Step 4: Determine the Ideal and the Negative-Ideal Solutions

We obtain the ideal solution A^* and the negative-ideal solution A^- from the weighted normalized matrix V. A^* is formed with the maximum values of benefit criteria and minimum values of cost criteria. In contrast, A^- is formed with the minimum values of benefit criteria and maximum values of cost criteria. The formations of A^* and A^- from the weighted normalized decision matrix are depicted in Table 3.3.

Hence, A^* and A^- are expressed as follows:

$$A^* = \left\{ v_1^*, v_2^*, v_3^* \right\}$$
$$= \{0.184, \ 0.184, \ 0.149\}$$

$$A^- = \left\{ v_1^-, v_2^-, v_3^- \right\}$$
$$= \{0.144, \ 0.156, \ 0.186\}$$

Step 5: Calculate the Separation Measure

In this step, the separation distances from the ideal (S_{i^*}) and the negative-ideal (S_{i-}) solutions for all alternatives are calculated. Tables 3.4 and 3.5 illustrate

TABLE 3.3

Formation of Ideal and Negative-Ideal Solutions

	Criteria	
Benefit		**Cost**
0.160	0.161	0.157
0.184*	0.156$^-$	0.186$^-$
0.144$^-$	0.184*	0.149*
0.176	0.165	0.172

TABLE 3.4

Calculation of Separation Measure from the Ideal Solution

	$(v_{i1} - v_1^*)^2$	$(v_{i2} - v_2^*)^2$	$(v_{i3} - v_3^*)^2$	$\sqrt{\sum_{j=1}^{3}(v_{ij} - v_j^*)^2}$
A_1	$(0.160 - 0.184)^2$	$(0.161 - 0.184)^2$	$(0.157 - 0.149)^2$	$S_{1^*} = 0.034$
A_2	$(0.184 - 0.184)^2$	$(0.156 - 0.184)^2$	$(0.186 - 0.149)^2$	$S_{2^*} = 0.046$
A_3	$(0.144 - 0.184)^2$	$(0.184 - 0.184)^2$	$(0.149 - 0.149)^2$	$S_{3^*} = 0.04$
A_4	$(0.176 - 0.184)^2$	$(0.165 - 0.184)^2$	$(0.172 - 0.149)^2$	$S_{4^*} = 0.03$

TABLE 3.5

Calculation of Separation Measure from the Negative-Ideal Solution

	$(v_{i1} - v_1^-)^2$	$(v_{i2} - v_2^-)^2$	$(v_{i3} - v_3^-)^2$	$\sqrt{\sum_{j=1}^{3}(v_{ij} - v_j^-)^2}$
A_1	$(0.160 - 0.144)^2$	$(0.161 - 0.156)^2$	$(0.157 - 0.186)^2$	$S_{1^-} = 0.033$
A_2	$(0.184 - 0.144)^2$	$(0.156 - 0.156)^2$	$(0.186 - 0.186)^2$	$S_{2^-} = 0.04$
A_3	$(0.144 - 0.144)^2$	$(0.184 - 0.156)^2$	$(0.149 - 0.186)^2$	$S_{3^-} = 0.046$
A_4	$(0.176 - 0.144)^2$	$(0.165 - 0.156)^2$	$(0.172 - 0.186)^2$	$S_{4^-} = 0.036$

the estimation of separation distances for all four alternatives from ideal and negative-ideal solutions, respectively. Figure 3.2 depicts the separation distances of all four alternatives from the ideal solution and the negative ideal solution.

Step 6: Calculate the Relative Closeness to the Ideal Solution

In order to determine the relative closeness to the ideal solution, we calculate $S_{i^-} / (S_{i^*} + S_{i^-})$ for each alternative. Table 3.6 displays the values of the relative closeness to the ideal solution for all four alternatives.

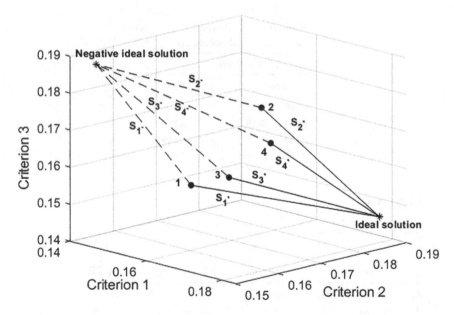

FIGURE 3.2
Separation measures of all four alternatives.

TABLE 3.6

Relative Closeness to the Ideal Solution

Alternatives	S_{i^*}	S_{i^-}	$C_{i^*} = \dfrac{S_{i^-}}{S_{i^*} + S_{i^-}}$	Rank
A_1	0.034	0.033	$C_{1^*} = 0.491$	3
A_2	0.046	0.04	$C_{2^*} = 0.463$	4
A_3	0.04	0.046	$C_{3^*} = 0.537$	2
A_4	0.03	0.036	$C_{4^*} = 0.544$	1

Step 7: Rank the Preference Order

Finally, we determine the preference order by arranging the alternatives in the descending order of C_{i^*}. Thus, the ranks for the alternatives in the selection problem using TOPSIS emerge as A_4, A_3, A_1, and A_2, which is shown in the last column of Table 3.6.

The MATLAB® coding for the aforesaid MCDM problem using TOPSIS is given in Section 3.7.

3.4 Application of TOPSIS in Textiles

Some applications of TOPSIS have been reported in the domain of textile engineering. Majumdar et al. (2005) used an analytic hierarchy process (AHP)-TOPSIS hybrid method of MCDM for determination of the quality value of cotton fiber. The ranking of cotton fibers attained by this AHP-TOPSIS hybrid method showed significant agreement with the ranking of yarn tenacity. The accuracy of this method was found to be higher than the existing methods. In another work, Majumdar et al. (2006) categorized cotton bales into different groups based on their TOPSIS values, and then bale laydown was formed using a frequency relative picking algorithm. The TOPSIS method was found to exert better control over the between laydown variances of cotton fiber properties as compared to the established spinning consistency index (SCI) method. Ghosh and Das (2013) used the TOPSIS method for grading jute fibers. Six quality parameters of jute fibers, viz., strength, defect, root content, color, fineness, and bulk density, were regarded as the criteria for the grading of raw jute. The criteria such as strength, root content, and fineness were measured experimentally, whereas defect, color, and bulk density were determined subjectively. The relative weights for different criteria were estimated using AHP. The 10 Tossa jute lots were ranked based on the relative closeness value determined by the TOPSIS method. The ranking of jute attained by this method showed significant agreement with the ranking based on the market price fixed by the local traders. Alam and Ghosh (2013) applied the TOPSIS method to evaluate the thermal comfort index of cotton fabrics by considering four criteria of fabric parameters, such as cover, thickness, areal density, and porosity. The ranking of fabrics on account of thermal comfort obtained by the TOPSIS method showed a reasonable degree of agreement with the ranking based on thermal resistance value. Majumdar et al. (2010) used TOPSIS for the selection of a suitable navel out of 10 navels for rotor spinning. Moghassem (2010) used the TOPSIS method to select the appropriate doffing tube and its adjustment for 30's Ne rotor yarn spun in order to raise the efficiency of the weft knitting machine. Shyjith et al. (2008) used AHP and TOPSIS methods for selecting an optimum maintenance strategy for a textile industry.

3.4.1 Selection of Handloom Fabrics for Summer Clothing Using TOPSIS

Mitra et al. (2015) used the AHP and multiplicative AHP (MAHP) of MCDM for ranking handloom cotton fabrics in terms of their overall quality value considering their applicability as summer clothing materials. Altogether, 25 plain woven 100% cotton handloom fabrics were manufactured using semiautomatic handloom. During sample manufacturing, fabric constructional parameters (ends per inch, picks per inch, warp count, and weft count) were varied as much as possible so that the samples cover a wide range of variability. All fabric samples were subjected to the testing of air permeability, drape coefficient, and thermal resistance.

Air permeability tests for the fabric specimens were conducted according to ASTM standard D737-96 on SDL Air Permeability Tester. Ten readings were taken for each fabric sample at random places without cutting the specimen, with an exposed area of 4 cm^2 under a pressure head of 10 mm of water column. The Cusick drape tester was used to measure the fabric drape coefficient. The higher the drape coefficient, the stiffer is the fabric—that is, the lower is the drapability. Thermal resistance of the fabric specimens was measured using an Alambeta instrument. In this instrument, the fabric is kept between the hot and cold plates. The hot plate comes in contact with the fabric sample at a pressure of 200 Pa. The test results of handloom fabric properties are shown in Table 3.7.

In this chapter, the TOPSIS method of MCDM has been implemented for developing an index of the quality of handloom fabrics, which would serve as summer clothing materials using the air permeability, drape, and thermal resistance results of 25 handloom cotton fabrics as given in Table 3.7. Figure 3.3 shows the hierarchical structure of this MCDM problem. With respect to summer application, air permeability was considered as the benefit criterion, whereas drape coefficient and thermal resistance were considered as the cost or negative (lower value is better) criteria. At first the relative

TABLE 3.7

Experimental Results of Fabric Properties

Fabric Code	Air Permeability $(cm^3 \cdot s^{-1} \cdot cm^{-2})$	Drape Coefficient (%)	Thermal Resistance $(m^2 \cdot K \cdot W^{-1} \times 10^{-3})$
A_1	243.420	65.900	11.000
A_2	217.92	67.620	11.130
A_3	244.14	68.280	12.010
A_4	216.45	67.250	11.680
A_5	218.56	66	12.67
A_6	241.52	80.09	12.05
A_7	224.44	75.76	14.17
A_8	214.23	63.34	13.16
A_9	253.97	70.04	15.98
A_{10}	241.38	76.24	7.97
A_{11}	246.62	79.53	13.66
A_{12}	214.45	78.43	10.9
A_{13}	242.45	64.81	12.24
A_{14}	238.67	68.76	12.04
A_{15}	241.26	65.39	12.44
A_{16}	246.72	65.08	24.65
A_{17}	236.61	67.01	12.01
A_{18}	258.5	70.97	13.27
A_{19}	258.5	66.65	13.69
A_{20}	214.69	68.09	13.62
A_{21}	215	64.72	12.54
A_{22}	217.62	68.31	12.22
A_{23}	244.5	80	14.22
A_{24}	236.54	65.48	13.85
A_{25}	239.28	77.94	12.08

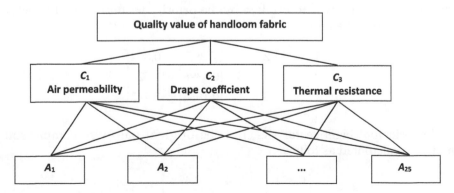

FIGURE 3.3
Hierarchical structure of the TOPSIS problem.

TABLE 3.8

Pair-Wise Comparison Matrix

Criteria	Drape Coefficient	Air Permeability	Thermal Resistance	Geometric Mean (GM)	Normalized GM or Relative Weights
Drape coefficient	1	1/2	2	1	0.297
Air permeability	2	1	3	1.817	0.5396
Thermal resistance	1/2	1/3	1	0.55	0.1634

weights of these three criteria were determined. For this purpose, the AHP method was used, which is discussed in the following lines.

With respect to the overall objective of the problem, the pair-wise comparison matrix of three criteria is given in Table 3.8. Here the comparisons were made according to Saaty's nine-point scale as discussed in the preceding chapter (Saaty, 1980). The scores given in Table 3.8 represent the perception of the decision maker about the relative importance of the three fabric parameters. These scores can vary from one decision maker to another and also with the intended use of the fabrics. It can be inferred from Table 3.8 that air permeability has moderate dominance over thermal resistance. However, the dominance of air permeability over the drape coefficient and the dominance of the drape coefficient over thermal resistance were somewhat between equal and moderate. The relative weights of criteria are shown in the last column of Table 3.8. It is observed that air permeability has the most dominant influence on the overall quality of the handloom fabrics with a relative weight of 0.5396. This seems to be rational as higher air permeability will ensure greater comfort to the wearer in the summer season. The relative weights of the drape coefficient and the thermal resistance are 0.2970 and 0.1634, respectively. For the measurement of consistency of judgment, the original pair-wise comparison matrix was multiplied by the priority vector or weight vector to get the product as follows:

$$\begin{bmatrix} 1 & 1/2 & 2 \\ 2 & 1 & 3 \\ 1/2 & 1/3 & 1 \end{bmatrix} \times \begin{bmatrix} 0.297 \\ 0.5396 \\ 0.1634 \end{bmatrix} = \begin{bmatrix} 0.894 \\ 1.624 \\ 0.492 \end{bmatrix}$$

The principal eigenvector (λ_{max}) of the original pair-wise comparison matrix is calculated as follows:

$$\lambda_{max} = \frac{\left(\dfrac{0.894}{0.297} + \dfrac{1.624}{0.5396} + \dfrac{0.492}{0.1634} \right)}{3} = 3.0092$$

To check the consistency in a pair-wise comparison, the consistency index (*CI*) and consistency ratio (*CR*) are calculated from the following equations:

$$CI = \frac{3.0092 - 3}{3 - 1} = 0.0046$$

$$CR = \frac{CI}{RCI} = \frac{0.0046}{0.58} = 0.0079 < 0.1$$

where *RCI* is the random consistency index, and for three criteria, its value is 0.58. The judgment is considered to be consistent since the calculated value of *CR* is less than 0.1.

After determining the weights of the decision criteria, the steps of the TOPSIS method are followed as explained earlier. Tables 3.9 and 3.10,

TABLE 3.9

Normalized Values of Fabric Properties

Alternatives	Air Permeability	Drape Coefficient	Thermal Resistance
A_1	0.207	0.188	0.165
A_2	0.185	0.192	0.167
A_3	0.208	0.194	0.181
A_4	0.184	0.191	0.176
A_5	0.186	0.188	0.190
A_6	0.205	0.228	0.181
A_7	0.191	0.216	0.213
A_8	0.182	0.180	0.198
A_9	0.216	0.199	0.240
A_{10}	0.205	0.217	0.120
A_{11}	0.210	0.226	0.205
A_{12}	0.182	0.223	0.164
A_{13}	0.206	0.184	0.184
A_{14}	0.203	0.196	0.181
A_{15}	0.205	0.186	0.187
A_{16}	0.210	0.185	0.371
A_{17}	0.201	0.191	0.181
A_{18}	0.220	0.202	0.199
A_{19}	0.220	0.190	0.206
A_{20}	0.183	0.194	0.205
A_{21}	0.183	0.184	0.188
A_{22}	0.185	0.194	0.184
A_{23}	0.208	0.228	0.214
A_{24}	0.201	0.186	0.208
A_{25}	0.204	0.222	0.182

TABLE 3.10

Weighted Normalized Values of Fabric Properties

Alternatives	Air Permeability	Drape Coefficient	Thermal Resistance
A_1	0.112	0.056	0.027
A_2	0.100	0.057	0.027
A_3	0.112	0.058	0.029
A_4	0.099	0.057	0.029
A_5	0.100	0.056	0.031
A_6	0.111	0.068⁻	0.030
A_7	0.103	0.064	0.035
A_8	**0.098⁻**	**0.054***	0.032
A_9	0.117	0.059	0.039
A_{10}	0.111	0.064	**0.020***
A_{11}	0.113	0.067	0.034
A_{12}	**0.098⁻**	0.066	0.027
A_{13}	0.111	0.055	0.030
A_{14}	0.110	0.058	0.030
A_{15}	0.111	0.055	0.031
A_{16}	0.113	0.055	**0.061⁻**
A_{17}	0.109	0.057	0.029
A_{18}	**0.119***	0.060	0.033
A_{19}	**0.119***	0.056	0.034
A_{20}	0.099	0.058	0.033
A_{21}	0.099	0.055	0.031
A_{22}	0.100	0.058	0.030
A_{23}	0.112	**0.068⁻**	0.035
A_{24}	0.109	0.055	0.034
A_{25}	0.110	0.066	0.030

respectively, show the normalized and weighted normalized values of three criteria for all alternatives.

The ideal solution A^* and negative ideal solution A^- are marked in Table 3.10. The values of A^* and A^- are thus obtained as follows:

$$A^* = \{0.119, 0.054, 0.020\}$$

$$A^- = \{0.098, 0.068, 0.061\}$$

The separation measures from the ideal (S_{i^*}) and the negative-ideal (S_{i^-}) solutions for all alternatives are calculated. In the next step, the relative closeness to the ideal solution (C_{i^*}) for each alternative is measured. This relative closeness to the ideal solution can be considered as the "handloom

TABLE 3.11

Relative Closeness to Ideal Solution and Rank of Alternatives

Alternatives	S_{i^*}	S_{i^-}	$C_{i^*} = S_{i^-}/(S_{i^*}+S_{i^-})$	TOPSIS Rank	AHP Rank
A_1	0.010	0.038	0.785	1	3
A_2	0.020	0.035	0.630	13	13
A_3	0.013	0.035	0.737	3	7
A_4	0.022	0.034	0.609	14	19
A_5	0.022	0.032	0.594	17	17
A_6	0.019	0.033	0.637	11	15
A_7	0.024	0.026	0.522	24	24
A_8	0.024	0.032	0.568	22	20
A_9	0.021	0.029	0.587	18	9
A_{10}	0.013	0.043	0.762	2	2
A_{11}	0.020	0.031	0.603	15	16
A_{12}	0.025	0.034	0.575	20	25
A_{13}	0.013	0.036	0.734	4	4
A_{14}	0.014	0.034	0.706	8	10
A_{15}	0.014	0.035	0.718	5	6
A_{16}	0.041	0.020	0.321	25	12
A_{17}	0.014	0.035	0.705	9	8
A_{18}	0.015	0.035	0.709	7	5
A_{19}	0.014	0.036	0.713	6	1
A_{20}	0.025	0.029	0.539	23	23
A_{21}	0.023	0.032	0.586	19	18
A_{22}	0.022	0.032	0.595	16	22
A_{23}	0.022	0.029	0.572	21	21
A_{24}	0.018	0.031	0.636	12	11
A_{25}	0.018	0.033	0.644	10	14

fabric selection index." Finally, the alternatives are ranked by arranging the values of C_{i^*} in descending order.

Table 3.11 depicts the values of $S_{i^*}, S_{i^-}, C_{i^*}$ and rank for all alternatives. These alternatives are also ranked in accordance with the AHP method of MCDM (Mitra et al. 2015), which is shown in the last column of Table 3.11. The rank correlation coefficient (R_s) is determined between the TOPSIS and AHP systems of MCDM using the following expression:

$$R_s = 1 - \frac{6\sum d_a^2}{n(n^2-1)} \tag{3.13}$$

where d_a is the absolute difference between two rankings, and n is the total number of alternatives. The rank correlation coefficient is obtained as 0.835, which shows a high degree of agreement between the two methods of MCDM.

3.5 Fuzzy-TOPSIS Method

In many circumstances, crisp data are insufficient to model real-life situations, because human assessments are often vague, which cannot be estimated with an exact numerical value (Zadeh, 1965). If such a situation arises in a MCDM problem, the ratings and weights of the criteria may be assessed by means of linguistic variables. Chen (2000) extended the concept of TOPSIS to develop a methodology for solving multiperson multicriteria decision-making problems in a fuzzy environment. Considering the fuzziness in the decision data and group decision-making process, linguistic variables are used to assess the weights of all criteria and the ratings of each alternative with respect to each criterion, and accordingly, the decision matrix is converted into a fuzzy decision matrix. Similar to the concept of TOPSIS, Chen (2000) defined the fuzzy positive ideal solution (FPIS) and the fuzzy negative ideal solution (FNIS). Consequently, using the vertex method, the distances of each alternative from FPIS as well as FNIS are calculated. Finally, a closeness coefficient of each alternative is determined for the purpose of ranking in preferential order. The higher value of the closeness coefficient indicates that an alternative is closer to FPIS and farther from FNIS simultaneously.

The fuzzy-TOPSIS method proposed by Chen (2000) is suitable for solving the group decision-making problem under a fuzzy environment where the weights of different criteria and the ratings of the qualitative criteria are considered as linguistic variables. These linguistic variables can be expressed in positive triangular fuzzy numbers as shown in Tables 3.12 and 3.13.

Suppose that a decision group has K persons, then the importance of the criteria and the rating of alternatives with respect to each criterion can be calculated as follows:

$$\tilde{x}_{ij} = \frac{1}{K}\left[\tilde{x}_{ij}^1(+)\tilde{x}_{ij}^2(+)\cdots(+)\tilde{x}_{ij}^K\right] \tag{3.14}$$

TABLE 3.12

Linguistic Variables for the Importance Weight of Each Criterion

Linguistic Variables	Triangular Fuzzy Numbers
Very low (VL)	(0, 0, 0.1)
Low (L)	(0, 0.1, 0.3)
Medium low (ML)	(0.1, 0.3, 0.5)
Medium (M)	(0.3, 0.5, 0.7)
Medium high (MH)	(0.5, 0.7, 0.9)
High (H)	(0.7, 0.9, 1)
Very high (VH)	(0.9, 1, 1)

TABLE 3.13

Linguistic Variables for the Ratings

Linguistic Variables	Triangular Fuzzy Numbers
Very poor (VP)	$(0, 0, 1)$
Poor (P)	$(0, 1, 3)$
Medium poor (MP)	$(1, 3, 5)$
Fair (F)	$(3, 5, 7)$
Medium good (MG)	$(5, 7, 9)$
Good (G)	$(7, 9, 10)$
Very good (VG)	$(9, 10, 10)$

$$\tilde{w}_j = \frac{1}{K}\left[\tilde{w}_j^1(+)\tilde{w}_j^2(+)\cdots(+)\tilde{w}_j^K\right] \tag{3.15}$$

where $\tilde{x}_{ij}^K$ and $\tilde{w}_j^K$ are the alternative rating and the criteria weight, respectively, given by the *Kth* decision maker. A fuzzy multicriteria group decision-making problem can be expressed in the matrix format as follows:

$$\tilde{D} = \begin{bmatrix} \tilde{x}_{11} & \tilde{x}_{12} & \cdots & \tilde{x}_{1n} \\ \tilde{x}_{21} & \tilde{x}_{22} & \cdots & \tilde{x}_{2n} \\ \cdot & \cdot & & \cdot \\ \cdot & \cdot & \cdots & \cdot \\ \cdot & \cdot & & \cdot \\ \tilde{x}_{m1} & \tilde{x}_{m2} & \cdots & \tilde{x}_{mn} \end{bmatrix} \tag{3.16}$$

$$\tilde{W} = \left[\tilde{w}_1, \tilde{w}_2, \ldots, \tilde{w}_n\right] \tag{3.17}$$

where $\tilde{x}_{ij}, \forall i, j$ and $\tilde{w}_j, j = 1, 2, \ldots, n$ are the linguistic variables, which can be described by triangular fuzzy numbers, $\tilde{x}_{ij} = (a_{ij}, b_{ij}, c_{ij})$ and $\tilde{w}_j = (w_{j1}, w_{j2}, w_{j3})$.

If B and C denote the set of benefit and cost criteria, respectively, the normalized fuzzy decision matrix is represented as follows:

$$\tilde{R} = \left[\tilde{r}_{ij}\right]_{m \times n} \tag{3.18}$$

where $\tilde{r}_{ij}$ can be expressed as

$$\tilde{r}_{ij} = \left(\frac{a_{ij}}{c_j^*}, \frac{b_{ij}}{c_j^*}, \frac{c_{ij}}{c_j^*}\right), j \in B \tag{3.19}$$

where

$$c_j^* = \max_i c_{ij}, \, j \in B \tag{3.20}$$

and

$$\tilde{r}_{ij} = \left(\frac{a_j^-}{c_{ij}}, \frac{a_j^-}{b_{ij}}, \frac{a_j^-}{a_{ij}} \right), \, j \in C \tag{3.21}$$

where

$$a_j^- = \min_i a_{ij}, \, j \in C \tag{3.22}$$

The weighted normalized fuzzy decision matrix can be expressed as

$$\tilde{V} = \left[\tilde{v}_{ij} \right]_{m \times n}, \, i = 1, 2, \ldots m; \, j = 1, 2, \ldots n \tag{3.23}$$

where

$$\tilde{v}_{ij} = \tilde{r}_{ij}(.)\tilde{w}_j \tag{3.24}$$

The FPIS (A^*) and FNIS (A^-) are defined as

$$A^* = \left(\tilde{v}_1^*, \tilde{v}_2^*, \ldots, \tilde{v}_n^* \right) \tag{3.25}$$

$$A^- = \left(\tilde{v}_1^-, \tilde{v}_2^-, \ldots, \tilde{v}_n^- \right) \tag{3.26}$$

where $\tilde{v}_j^*(1,1,1)$ and $\tilde{v}_j^- = (0,0,0)$; $j = 1, 2, \ldots, n$. The separation distances from each alternative from A^* and A^- can be estimated as

$$d_i^* = \sum_{j=1}^{n} d\left(\tilde{v}_{ij}, \tilde{v}_j^* \right), \, i = 1, 2, \ldots, m \tag{3.27}$$

$$d_i^- = \sum_{j=1}^{n} d\left(\tilde{v}_{ij}, \tilde{v}_j^- \right), \, i = 1, 2, \ldots, m \tag{3.28}$$

where $d(.,.)$ is the distance measurement between two fuzzy numbers. For example, if $\tilde{m} = (m_1, m_2, m_3)$ and $\tilde{n} = (n_1, n_2, n_3)$ are the two fuzzy numbers, then the distance between $\tilde{m}$ and $\tilde{n}$ can be calculated as

$$d(\tilde{m},\tilde{n}) = \sqrt{\frac{1}{3}\left[(m_1 - n_1)^2 + (m_2 - n_2)^2 + (m_3 - n_3)^2\right]}$$

The closeness coefficient of each alternative is estimated as

$$CC_i = \frac{d_i^-}{d_i^* + d_i^-}, i = 1, 2, \ldots, m \tag{3.29}$$

As CC_i approaches one, an alternative A_i is said to be closer to FPIS (A^*) and further from FNIS (A^-). The ranking order of the alternatives is determined according to the closeness coefficient.

In brief, various steps involved in a fuzzy-TOPSIS MCDM method may be summarized as follows:

- *Step 1*: Form a committee of decision makers, and identify the evaluation criteria.
- *Step 2*: Select the appropriate linguistic variables for the importance weight of the criteria and the linguistic ratings for alternatives with respect to criteria.
- *Step 3*: Aggregate the weight of criteria to get the aggregated fuzzy weight $\tilde{w}_j$ of criterion C_j, and pool the decision makers' judgments to obtain the aggregated fuzzy rating $\tilde{x}_{ij}$ of alternative A_i under criterion C_j.
- *Step 4*: Construct the fuzzy decision matrix and the normalized fuzzy decision matrix.
- *Step 5*: Construct the weighted normalized fuzzy decision matrix.
- *Step 6*: Determine the FPIS and FNIS.
- *Step 7*: Estimate the distance of each alternative from FPIS and FNIS, respectively.
- *Step 8*: Estimate the closeness coefficient of each alterative.
- *Step 9*: Determine the ranking order of each alternative, according to the closeness coefficient.

There are only limited reports available on the application of fuzzy-TOPSIS in textiles. Yayla et al. (2012) used the fuzzy TOPSIS method to select the most appropriate supplier of garments. In the solution of the problem, five criteria such as quality, delivery time, cost, flexibility, and geographical location were considered. The decision-making group was composed of three experts. Nazam et al. (2015) formulated a fuzzy AHP-TOPSIS framework for the risk assessment of green supply chain implementation in the textile industry.

3.6 Step-by-Step Working Principles of Fuzzy-TOPSIS

The working principles of the fuzzy-TOPSIS MCDM method have been discussed with an example of the grading of silk cocoons. Three lots of mulberry multibivoltine cocoons A_1, A_2, and A_3 are collected from the silk cocoon market. A committee of three decision makers D_1, D_2, and D_3 has been formed for grading the cocoon and selecting the most suitable lot. Four benefit criteria of cocoons are considered, such as flawless cocoon (%), weight, size, and resilience. The hierarchical structure of this multiperson MCDM problem is displayed in Figure 3.4. The fuzzy-TOPSIS method proposed by Chen (2000) has been applied to this cocoon grading problem, and the step-by-step computational procedure is summarized as follows.

Step 1

Three experts use the linguistic weighting variables of Table 3.12 for assessing the importance of the criteria that are presented in Table 3.14.

Step 2

The experts use the linguistic rating variables of Table 3.13 for evaluating the rating of alternatives with respect to each criterion as presented in Table 3.15.

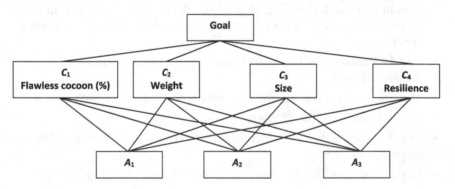

FIGURE 3.4
Hierarchical structure of the fuzzy-TOPSIS problem.

TABLE 3.14

The Importance Weight of the Criteria Assessed by Three Experts

	D_1	D_2	D_3
C_1 (Flawless cocoon %)	VH	VH	H
C_2 (Weight)	M	M	VH
C_3 (Size)	L	M	L
C_4 (Resilience)	H	VH	VH

TABLE 3.15

Ratings of the Three Alternatives by Three Decision Makers for All Four Criteria

Criteria	Alternatives	Decision Makers		
		D_1	D_2	D_3
C_1	A_1	P	F	MP
	A_2	MG	MG	G
	A_3	G	MG	VG
C_2	A_1	P	MP	F
	A_2	P	F	MG
	A_3	P	MP	F
C_3	A_1	G	MG	MG
	A_2	MG	F	MP
	A_3	MG	F	MP
C_4	A_1	G	G	G
	A_2	F	MG	F
	A_3	VG	VG	VG

TABLE 3.16

Fuzzy Decision Matrix and Fuzzy Weights of All Four Criteria

	C_1	C_2	C_3	C_4
A_1	(1.33, 3, 5)	(1.33, 3, 5)	(5.67, 7.67, 9.33)	(7, 9, 10)
A_2	(5.67, 7.67, 9.33)	(2.67, 4.33, 6.33)	(3, 5, 7)	(3.67, 5.67, 7.67)
A_3	(7, 8.67, 9.67)	(1.33, 3, 5)	(3, 5, 7)	(9, 10, 10)
Weight	(0.83, 0.97, 1)	(0.5, 0.67, 0.8)	(0.1, 0.23, 0.43)	(0.83, 0.97, 1)

Step 3

The linguistic evaluations of Tables 3.14 and 3.15 are converted into triangular fuzzy numbers to construct the fuzzy decision matrix and to determine the fuzzy weight of each criterion as shown in Table 3.16. For example, the ratings of the first cocoon lot A_1 with respect to the first criterion C_1 given by three experts are poor (P), fair (F), and medium poor (MP), respectively. Furthermore, for the first criterion, three experts have given very high (VH), very high (VH), and high (H) importance weights, respectively. Hence, using Equation 3.14, the aggregated fuzzy rating $\tilde{x}_{11}$ of alternative A_1 with respect to criterion C_1 can be calculated as

$$\tilde{x}_{11} = \frac{1}{3}\left[\tilde{x}_{11}^1(+)\tilde{x}_{11}^2(+)\tilde{x}_{11}^3\right]$$

$$= \frac{1}{3}[(0,1,3) + (3,5,7) + (1,3,5)]$$

$$= \frac{1}{3}[(4,9,15)]$$

$$= (1.33, 3, 5)$$

TABLE 3.17

Fuzzy Normalized Decision Matrix

	C_1	C_2	C_3	C_4
A_1	(0.14, 0.31, 0.52)	(0.21, 0.47, 0.79)	(0.61, 0.82, 1)	(0.7, 0.9, 1)
A_2	(0.59, 0.79, 0.97)	(0.42, 0.68, 1)	(0.32, 0.54, 0.75)	(0.37, 0.57, 0.77)
A_3	(0.72, 0.9, 1)	(0.21, 0.47, 0.79)	(0.32, 0.54, 0.75)	(0.9, 1, 1)

Similarly, using Equation 3.15, the aggregated weight of the first criterion C_1 can be calculated as

$$
\begin{aligned}
\tilde{w}_1 &= \frac{1}{3}\left[\tilde{w}_1^1(+)\tilde{w}_1^2(+)\tilde{w}_1^3\right] \\
&= \frac{1}{3}[(0.9,1,1)+(0.9,1,1)+(0.7,0.9,1)] \\
&= \frac{1}{3}[(2.5,2.9,3)] \\
&= (0.83,0.97,1)
\end{aligned}
$$

Step 4

The normalized fuzzy decision matrix is constructed as shown in Table 3.17. As an example, for the first criterion C_1, using Equation 3.20 we have

$$
c_1^* = \max_i c_{i1} = 9.67
$$

Hence, using Equation 3.19, the normalized triangular fuzzy numbers of alternative A_1 for first criterion C_1 can be calculated as

$$
\begin{aligned}
\tilde{r}_{11} &= \left(\frac{a_{11}}{c_1^*}, \frac{b_{11}}{c_1^*}, \frac{c_{11}}{c_1^*}\right) \\
&= \left(\frac{1.33}{9.67}, \frac{3}{9.67}, \frac{5}{9.67}\right) \\
&= (0.14,0.31,0.52)
\end{aligned}
$$

Step 5

The weighted normalized fuzzy decision matrix is constructed as shown in Table 3.18. For example, using Equation 3.24, the weighted normalized triangular fuzzy numbers of alternative A_1 for first criterion C_1 can be estimated as

TABLE 3.18

Fuzzy Weighted Normalized Decision Matrix

	C_1	C_2	C_3	C_4
A_1	(0.11, 0.3, 0.52)	(0.11, 0.32, 0.63)	(0.06, 0.19, 0.43)	(0.58, 0.87, 1)
A_2	(0.49, 0.77, 0.97)	(0.21, 0.46, 0.8)	(0.03, 0.12, 0.33)	(0.31, 0.55, 0.77)
A_3	(0.6, 0.87, 1)	(0.11, 0.32, 0.63)	(0.03, 0.12, 0.33)	(0.75, 0.97, 1)

$$\tilde{v}_{11} = \tilde{r}_{11}(.)\tilde{w}_1$$
$$= (0.14, 0.31, 0.52).(0.83, 0.97, 1)$$
$$= (0.14 \times 0.83, 0.31 \times 0.97, 0.52 \times 1)$$
$$= (0.11, 0.3, 0.52)$$

Step 6

The FPIS and FNIS are determined as follows:

$$A^* = [(1,1,1),(1,1,1),(1,1,1),(1,1,1)],$$

$$A^- = [(0,0,0),(0,0,0),(0,0,0),(0,0,0)].$$

Step 7

In this step, the distances of each alternative from the FPIS and FNIS are calculated. Using Equations 3.27 and 3.28, the distances from the FPIS and FNIS for alternative A_1 are calculated as

$$d_1^* = \sum_{j=1}^{4} d\left(\tilde{v}_{1j}, \tilde{v}_j^*\right)$$

$$= \sqrt{\frac{1}{3}[(0.11-1)^2 + (0.3-1)^2 + (0.52-1)^2]} + \sqrt{\frac{1}{3}[(0.11-1)^2 + (0.32-1)^2 + (0.63-1)^2]}$$

$$+ \sqrt{\frac{1}{3}[(0.06-1)^2 + (0.19-1)^2 + (0.43-1)^2]} + \sqrt{\frac{1}{3}[(0.58-1)^2 + (0.87-1)^2 + (1-1)^2]}$$

$$= 2.43$$

$$d_1^- = \sum_{j=1}^{4} d\left(\tilde{v}_{1j}, \tilde{v}_j^-\right)$$

$$= \sqrt{\frac{1}{3}[(0.11-0)^2 + (0.3-0)^2 + (0.52-0)^2]} + \sqrt{\frac{1}{3}[(0.11-0)^2 + (0.32-0)^2 + (0.63-0)^2]}$$

$$+ \sqrt{\frac{1}{3}[(0.06-0)^2 + (0.19-0)^2 + (0.43-0)^2]} + \sqrt{\frac{1}{3}[(0.58-0)^2 + (0.87-0)^2 + (1-0)^2]}$$

$$= 1.88$$

TABLE 3.19

Distance from FPIS and FNIS and Closeness Coefficient

	d_i^*	d_i^-	$CC_i = d_i^- / (d_i^* + d_i^-)$	Rank
A_1	2.43	1.88	0.44	3
A_2	2.24	2.08	0.48	2
A_3	1.92	2.37	0.55	1

Similarly, the distances from the FPIS and FNIS for the other alternatives A_2 and A_3 can be calculated. Table 3.19 shows the distances from FPIS and FNIS for each alternative.

Step 8

In order to determine the closeness coefficient of each alternative, we calculate $d_i^- / (d_i^* + d_i^-)$ for each alternative. Table 3.19 shows the values of closeness coefficient for all three alternatives.

Step 9

Finally, we determine the ranking order by arranging the alternatives in the descending order of CC_i. Thus, the ranking order of the three alternatives is A_3, A_2, and A_1 as shown in the last column of Table 3.19.

The MATLAB® coding for the aforesaid multicriteria decision-making problem using fuzzy-TOPSIS is given in Section 3.7.

3.7 MATLAB® Coding

MATLAB® Coding for TOPSIS

```
clc
close all
clear all
format short g

% Step 1: Construct the decision matrix
% (Alternatives are in rows, criteria are in columns)
a = [100   35    55
     115   34    65
      90   40    52
     110   36    60];
[s1,s2] = size(a);
```

```
% Cost and benefit criteria
cc = 3;
bc = 1:2;

% Step 2: Calculate normalized decision matrix
sq = a.^2;
j = 1 : s2;
s = sqrt(sum(sq(:,j)));
for j = 1 : s2;
    X(:,j) = (a(:,j))/s(j);
end

% Step 3: Calculate the weighted normalized matrix
% Weight assigned to the criteria
w = [1/3 1/3 1/3];
W = diag(w);
y = X*W;

% Step 4: Determine the ideal and negative ideal solutions
% Ideal solutions
for j = bc
  ma(j) = max(y(:,j));
  ideal(j) = ma(j);
end
for j = cc
  mi(j) = min(y(:,j));
  nonideal(j) = mi(j);
end
I = [ideal(bc), nonideal(cc)];

% Negative ideal solutions
for j = bc
    mi_ni(j) = min(y(:,j));
    neg_ideal(j) = mi_ni(j);
end
for j=cc
    ma_ni(j) = max(y(:,j));
    neg_nonideal(j) = ma_ni(j);
end
NI = [neg_ideal(bc), neg_nonideal(cc)];

% Step 5: Calculation of the separation measure
% Ideal separation measure
for i = 1 : s1;
  for j = 1 : s2;
    z1(i,j) = (y(i,j)-I(j)).^2;
  end
end
for i = 1: s1;
    S1(i) = sqrt(sum(z1(i,:)));
end
```

```
% Negative ideal separation measure
for i = 1: s1;
  for j = 1: s2;
      z2(i,j) = (y(i,j) - NI(j)).^2;
    end
end
for i=1:s1
  S2(i) = sqrt(sum(z2(i,:)));
end

% Step 6: Calculation of relative closeness to the ideal solution
for i = 1 : s1
  C(i) = S2(i)./(S1(i) + S2(i));
end

% Step 7: Rank the preference order
alternative = 1 : s1;
Topsis_Score = [C']
[c, Index] = sort(C, 'descend');
[i, rank] = sort(Index, 'ascend');
[alternative' rank']
```

MATLAB® Coding for Fuzzy-TOPSIS

```
clc
close all
clear all
format short

VL=[0 0 0.1];
L=[0 0.1 0.3];
ML=[0.1 0.3 0.5];
M=[0.3 0.5 0.7];
MH=[0.5 0.7 0.9];
H=[0.7 0.9 1];
VH=[0.9 1 1];

VP=[0 0 1];
P=[0 1 3];
MP=[1 3 5];
F= [3 5 7];
MG=[5 7 9];
G=[7 9 10];
VG=[9 10 10];

for i=1:3
W1(:,i)=mean([VH(i) VH(i) H(i)]);
W2(:,i)=mean([M(i) M(i) VH(i)]);
W3(:,i)=mean([L(i) M(i) L(i)]);
W4(:,i)=mean([H(i) VH(i) VH(i)]);
end
```

```
for i=1:3
A1C1(:,i)=mean([P(i)  F(i)  MP(i)]);
A2C1(:,i)=mean([MG(i) MG(i) G(i)]);
A3C1(:,i)=mean([G(i)  MG(i) VG(i)]);
A1C2(:,i)=mean([P(i)  MP(i) F(i)]);
A2C2(:,i)=mean([P(i)  F(i)  MG(i)]);
A3C2(:,i)=mean([P(i)  MP(i) F(i)]);
A1C3(:,i)=mean([G(i)  MG(i) MG(i)]);
A2C3(:,i)=mean([MG(i) F(i)  MP(i)]);
A3C3(:,i)=mean([MG(i) F(i)  MP(i)]);
A1C4(:,i)=mean([G(i)  G(i)  G(i)]);
A2C4(:,i)=mean([F(i)  MG(i) F(i)]);
A3C4(:,i)=mean([VG(i) VG(i) VG(i)]);
end

m1=max([A1C1  A2C1 A3C1]);
m2=max([A1C2  A2C2 A3C2]);
m3=max([A1C3  A2C3 A3C3]);
m4=max([A1C4  A2C4 A3C4]);

A1C1N=A1C1./m1;
A2C1N=A2C1./m1;
A3C1N=A3C1./m1;
A1C2N=A1C2./m2;
A2C2N=A2C2./m2;
A3C2N=A3C2./m2;
A1C3N=A1C3./m3;
A2C3N=A2C3./m3;
A3C3N=A3C3./m3;
A1C4N=A1C4./m4;
A2C4N=A2C4./m4;
A3C4N=A3C4./m4;

A1C1W=A1C1N.*W1;
A2C1W=A2C1N.*W1;
A3C1W=A3C1N.*W1;
A1C2W=A1C2N.*W2;
A2C2W=A2C2N.*W2;
A3C2W=A3C2N.*W2;
A1C3W=A1C3N.*W3;
A2C3W=A2C3N.*W3;
A3C3W=A3C3N.*W3;
A1C4W=A1C4N.*W4;
A2C4W=A2C4N.*W4;
A3C4W=A3C4N.*W4;

A_p=ones(1,3);
A_m=zeros(1,3);

A1A_p = (sqrt(sum((A1C1W-A_p).^2))+sqrt(sum((A1C2W-A_p).^2)) ...
        + sqrt(sum((A1C3W-A_p).^2))+sqrt(sum((A1C4W-A_p).^2)))/sqrt(3)
A2A_p = (sqrt(sum((A2C1W-A_p).^2))+sqrt(sum((A2C2W-A_p).^2)) ...
        + sqrt(sum((A2C3W-A_p).^2))+sqrt(sum((A2C4W-A_p).^2)))/sqrt(3)
A3A_p = (sqrt(sum((A3C1W-A_p).^2))+sqrt(sum((A3C2W-A_p).^2)) ...
        + sqrt(sum((A3C3W-A_p).^2))+sqrt(sum((A3C4W-A_p).^2)))/sqrt(3)
```

```
A1A_m = (sqrt(sum((A1C1W-A_m).^2))+sqrt(sum((A1C2W-A_m).^2)) …
      + sqrt(sum((A1C3W-A_m).^2))+sqrt(sum((A1C4W-A_m).^2)))/sqrt(3)
A2A_m = (sqrt(sum((A2C1W-A_m).^2))+sqrt(sum((A2C2W-A_m).^2)) …
      + sqrt(sum((A2C3W-A_m).^2))+sqrt(sum((A2C4W-A_m).^2)))/sqrt(3)
A3A_m = (sqrt(sum((A3C1W-A_m).^2))+sqrt(sum((A3C2W-A_m).^2)) …
      + sqrt(sum((A3C3W-A_m).^2))+sqrt(sum((A3C4W-A_m).^2)))/sqrt(3)

S1=[A1A_p A2A_p A3A_p];
S2=[A1A_m A2A_m A3A_m];

for i=1:length(S1)
  C(i)=S2(i)./(S1(i)+S2(i));
end

alternative=1:length(S1);
Topsis_Score=[C']
[c,Index]=sort(C,'descend');
[i,rank]=sort(Index,'ascend');
[alternative' rank']
```

3.8 Summary

TOPSIS and fuzzy-TOPSIS methods of MCDM are discussed in this chapter with the aid of step-by-step, worked out problems including the MATLAB® coding. A survey of literature on the application of TOPSIS and fuzzy-TOPSIS in various MCDM problems pertaining to textiles has been discussed. In addition, this chapter includes a detailed study on the selection of handloom fabrics for summer clothing using the TOPSIS method. An application of the fuzzy-TOPSIS method is also discussed with an example of a silk cocoon grading problem.

References

Alam, S. and Ghosh, A. 2013. Selection of cotton fabrics for optimal comfort properties using multi-criteria decision making. *Journal of Textiles and Apparel Technology Management*, 8(3), 1–8.

Chen, C. T. 2000. Extensions of the TOPSIS for group decision-making under fuzzy environment. *Fuzzy Sets and Systems*, 114, 1–9.

Ghosh, A. and Das, S. 2013. Raw jute grading by multi-criteria decision making method. *Journal of Natural Fibers*, 10, 136–146.

Hwang, C. L. and Yoon, K. 1981. *Multiple Attribute Decision Making: Methods and Applications*, Springer-Verlag, New York.

Majumdar, A., Kaplan, S. and Göktepe, Ö. 2010. Navel selection for rotor spinning denim fabrics using a multi-criteria decision making process. *Journal of Textile Institute*, 101(4), 304–309.

Majumdar, A., Sarkar, B. and Majumdar, P. K. 2005. Determination of quality value of cotton fibre using hybrid AHP-TOPSIS method of multi-criteria decision making. *Journal of the Textile Institute*, 96(5), 303–309.

Majumdar, A., Sarkar, B. and Majumdar, P. K. 2006. A new algorithm of cotton fibre selection and laydown using TOPSIS method of multicriteria decision making. *Indian Journal of Fibre and Textile Research*, 31, 248–255.

Mitra, A., Majumdar, A., Ghosh, A., Majumdar, P. K. and Bannerjee, D. 2015. Selection of handloom fabrics for summer clothing using multi-criteria decision making techniques. *Journal of Natural Fibers*, 12, 61–71.

Moghassem, A. R. 2010. Application of TOPSIS approach on parameters selection problem for rotor spinning machine. *Fibers and Polymers*, 11(4), 669–675.

Nazam, M., Xu, J., Tao, Z., Ahmed, J. and Hashim, M. 2015. A fuzzy AHP-TOPSIS framework for the risk assessment of green supply chain implementation in the textile industry. *International Journal of Supply and Operations Management*, 2(1), 548–568.

Saaty, T. L. 1980. *The Analytic Hierarchy Process*, McGraw-Hill, New York.

Shyjith, K., Ilangkumaran, M. and Kumanan, S. 2008. Multi-criteria decision-making approach to evaluate optimum maintenance strategy in textile industry. *Journal of Quality in Maintenance Engineering*, 14(4), 375–386.

Yayla, A. Y., Yildiz, A. and Özbek, A. 2012. Fuzzy TOPSIS method in supplier selection and application in the garment industry. *Fibres & Textiles in Eastern Europe*, 20(4), 20–23.

Zadeh, L. A. 1965. Fuzzy Sets. *Information and Control*, 8, 338–353.

4

Elimination and Choice Translating Reality (ELECTRE)

4.1 Introduction

The ELECTRE (elimination et choix traduisant la realité [elimination and choice translating reality]) method was introduced by Benayoun et al. (1966) in France. This method belongs to the family of outranking techniques of MCDM. The outranking technique of MCDM is used to select which alternative is preferable, indifferent, or incomparable by comparing an alternative with another one under each criterion. The benefit of the outranking method is its ability to take purely ordinal scales into account, where indifference and preference thresholds can be considered when modeling the imperfect knowledge of data.

The ELECTRE method aims at building a binary outranking relation S, where aSb means "a is at least as good as b." Several versions of the ELECTRE method exist in the literature, such as ELECTRE I, II, III, IV, TRI, etc. These versions have been widely recognized as effective and efficient MCDM tools, with successful applications in different domains (Damaskos and Kalfakakou, 2005; Papadopoulos and Karagiannidis, 2008; Abedi et al., 2012; Sánchez-Lozano et al., 2014). This chapter deals only with the ELECTRE I method, which is simply known as ELECTRE.

4.2 ELECTRE Methodology

The ELECTRE method deals with the "outranking relations" by means of pairwise comparisons among alternatives under each one of the criteria separately. An interpretation of the outranking relationship of the two alternatives A_i and A_j, denoted by $A_i \rightarrow A_j$ as given by Roy (1991) is that even when the ith alternative does not dominate the jth alternative quantitatively, then the decision maker may still take the risk of regarding A_i as almost surely better than A_j. Alternatives are considered to be dominated if there is another alternative that surpasses them in one or more criteria and equals the remaining criteria (Triantaphyllou, 2000).

All criteria in different alternatives split into two different subsets, namely, the concordance set and the discordance set. The former is composed of the set of criteria for which alternative A_i outranks or dominates alternative A_j, and the latter is the complementary subset. In the process of successive evaluations of the outranking relations of the alternatives, the ELECTRE method obtains the concordance index, defined as the amount of evidence to support the conclusion that A_i dominates alternative A_j. The counterpart of the concordance index is the discordance index. In due course, the ELECTRE method yields a system of binary outranking relations between the alternatives. In this method, less favorable alternatives are basically eliminated, but sometime it is unable to identify the most preferred alternative. There are many versions of the ELECTRE method. The principle of the original version of the ELECTRE method is discussed in the following steps.

Step 1: Construct the Decision Matrix

This step constructs decision matrix D of m alternatives and n criteria with the information available regarding the MCDM problem as follows:

$$D = \begin{bmatrix} x_{11} & x_{12} & \cdots & x_{1n} \\ x_{21} & x_{22} & \cdots & x_{2n} \\ \cdots & \cdots & \cdots & \cdots \\ x_{m1} & x_{m2} & \cdots & x_{mn} \end{bmatrix} \tag{4.1}$$

where x_{ij} denotes the performance measure of the ith alternative in terms of the jth criterion.

Step 2: Construct the Normalized Decision Matrix

This step transforms the entries of the decision matrix into dimensionless comparable entries. An element r_{ij} of the normalized decision matrix R is estimated as follows:

$$r_{ij} = \frac{x_{ij}}{\sqrt{\sum_{k=1}^{m} x_{kj}^2}} \tag{4.2}$$

and

$$R = \begin{bmatrix} r_{11} & r_{12} & \cdots & r_{1n} \\ r_{21} & r_{22} & \cdots & r_{2n} \\ \cdots & \cdots & \cdots & \cdots \\ r_{m1} & r_{m2} & \cdots & r_{mn} \end{bmatrix} \tag{4.3}$$

Step 3: Construct the Weighted Normalized Matrix

Each one of the columns of the normalized decision matrix R is multiplied by the associated weight of importance of the corresponding decision criterion to obtain the weighted normalized matrix V as follows:

$$V = RW \tag{4.4}$$

where

$$W = \begin{bmatrix} w_1 & 0 & 0 & 0 & \cdots & 0 \\ 0 & w_2 & 0 & 0 & \cdots & 0 \\ \cdots & \cdots & \cdots & \cdots & \cdots & \cdots \\ 0 & 0 & 0 & 0 & \cdots & w_n \end{bmatrix} \tag{4.5}$$

and $\sum w_i = 1$,

The expression of V thus becomes

$$V = \begin{bmatrix} w_1 r_{11} & w_2 r_{12} & w_3 r_{13} & \cdots & w_n r_{1n} \\ w_1 r_{21} & w_2 r_{22} & w_3 r_{23} & \cdots & w_n r_{2n} \\ \cdot & & & & \cdot \\ \cdot & & & & \cdot \\ \cdot & & & & \cdot \\ w_1 r_{m1} & w_2 r_{m2} & w_3 r_{m3} & \cdots & w_n r_{mn} \end{bmatrix} \tag{4.6}$$

An element v_{ij} of the weighted normalized decision matrix is estimated as follows:

$$v_{ij} = w_j \cdot r_{ij} \tag{4.7}$$

It may be noted that the first three steps of the ELECTRE method are the same as those of the TOPSIS, and the difference between these two methods begins from the subsequent steps.

Step 4: Determine the Concordance and Discordance Sets

The concordance set C_{kl} of two alternatives A_k and A_l, where $m \geq k, l \geq 1$, is defined as the set of all the criteria for which A_k is preferred to A_l, that is, the following is true:

for profit criterion,

$$C_{kl} = \{j, v_{kj} \geq v_{lj}\}, \quad \text{for } j = 1,2,3,\ldots,n$$

and for cost criterion,

$$C_{kl} = \{j, v_{kj} < v_{lj}\}, \quad \text{for } j = 1,2,3,\ldots,n$$

The complementary subset is termed as discordance set D_{kl}, and it is denoted as follows:

for profit criterion,

$$D_{kl} = \{j, v_{kj} < v_{lj}\}, \quad \text{for } j = 1,2,3,\ldots,n$$

and for cost criterion,

$$D_{kl} = \{j, v_{kj} \geq v_{lj}\}, \quad \text{for } j = 1,2,3,\ldots,n$$

Step 5: Construct the Concordance and Discordance Matrices

The relative value of the elements in the concordance matrix C is estimated by means of the concordance index. The concordance index c_{kl} is the sum of the weights associated with the criteria contained in the concordance set, that is, the following is true:

$$c_{kl} = \sum_{j \in C_{kl}} w_j, \quad \text{for } j = 1,2,3,\ldots,n \tag{4.8}$$

The concordance index signifies the relative dominance of the alternative A_k with respect to the alternative A_l. The value of c_{kl} lies between 0 and 1. The concordance matrix C is expressed as follows:

$$C = \begin{vmatrix} - & c_{12} & \cdots & c_{1m} \\ c_{21} & - & \cdots & c_{2m} \\ \cdots & \cdots & \cdots & \cdots \\ c_{m1} & c_{m2} & \cdots & - \end{vmatrix} \tag{4.9}$$

where the entries of matrix C are not defined when $k = l$.

The discordance matrix D expresses the degree of weakness of alternative A_k with respect to a competing alternative A_l. The elements d_{kl} of the discordance matrix are defined as follows:

$$d_{kl} = \frac{\max\limits_{j \in D_{kl}} |v_{kj} - v_{lj}|}{\max\limits_{j} |v_{kj} - v_{lj}|} \tag{4.10}$$

The discordance matrix is expressed as follows:

$$D = \begin{vmatrix} - & d_{12} & \cdots & d_{1m} \\ d_{21} & - & \cdots & d_{2m} \\ \cdots & \cdots & \cdots & \cdots \\ d_{m1} & d_{m2} & \cdots & - \end{vmatrix} \qquad (4.11)$$

Similar to the C matrix, the entries of matrix D are not defined when $k = l$.

Step 6: Determine the Concordance and Discordance Dominance Matrices

The formation of the concordance dominance matrix depends on the threshold value for the concordance index. Alternative A_k will only have the possibility of dominating alternative A_l if its corresponding concordance index c_{kl} exceeds at least a certain threshold value $\underline{c}$, that is, if the following condition is true:

$$c_{kl} \geq \underline{c} \qquad (4.12)$$

The threshold value $\underline{c}$ can be estimated as the average concordance index as follows:

$$\underline{c} = \frac{1}{m(m-1)} \sum_{\substack{k=1 \\ k \neq l}}^{m} \sum_{\substack{l=1 \\ l \neq k}}^{m} c_{kl} \qquad (4.13)$$

On the basis of threshold value, the elements of the concordance dominance matrix F are determined as follows:

$$f_{kl} = 1, \quad \text{if } c_{kl} \geq \underline{c}$$

$$f_{kl} = 0, \quad \text{if } c_{kl} < \underline{c}$$

Similarly, the discordance dominance matrix G is defined by using a threshold value $\underline{d}$, which is defined as follows:

$$\underline{d} = \frac{1}{m(m-1)} \sum_{\substack{k=1 \\ k \neq l}}^{m} \sum_{\substack{l=1 \\ l \neq k}}^{m} d_{kl} \qquad (4.14)$$

The elements of the matrix G are determined as follows:

$$g_{kl} = 1, \quad \text{if } d_{kl} \geq \underline{d}$$

$$g_{kl} = 0, \quad \text{if } d_{kl} < \underline{d}$$

Step 7: Determine the Aggregate Dominant Matrix

The elements of the aggregate dominance matrix E are determined as follows:

$$e_{kl} = f_{kl} \times g_{kl} \qquad (4.15)$$

Step 8: Eliminate the Less Favorable Alternatives

From the aggregate dominance matrix E, a preference ordering of the alternatives can be drawn. For example, $e_{kl} = 1$ means that alternative A_k is preferred to alternative A_l by using both the concordance and discordance criteria. Therefore, if $e_{kl} = 1$, then A_k is strictly preferred to A_l. If $e_{kl} = 1$ and $e_{lk} = 1$, then A_k is indifferent to A_l. If $e_{kl} = 0$ and $e_{lk} = 0$, then A_k is incomparable to A_l. If any column of the matrix E has at least one element equal to 1, then this column is "ELECTREally" dominated by the corresponding row (Triantaphyllou, 2000). Thus, any column that has an element equal to one can be eliminated. The best alternative is the one that dominates all other alternatives.

4.3 Step-by-Step Working Principles of ELECTRE Method

In order to explain the working principles of the ELECTRE method in a step-by-step manner, the same MCDM problem as discussed in Chapter 3, Section 3.3 is considered. The decision matrix of the MCDM problem consists of four alternatives and three criteria as shown in Table 4.1. In this given problem, C_1 and C_2 are benefit criteria, and C_3 is the cost criterion.

The following steps are involved to explain the working principles of the ELECTRE method.

TABLE 4.1

Decision Matrix of the Problem

Criteria		C_1	C_2	C_3
Weights		1/3	1/3	1/3
Alternatives	A_1	100	35	55
	A_2	115	34	65
	A_3	90	40	52
	A_4	110	36	60

Step 1: Construct the Decision Matrix

The decision matrix can be represented as follows:

$$D = \begin{bmatrix} 100 & 35 & 55 \\ 115 & 34 & 65 \\ 90 & 40 & 52 \\ 110 & 36 & 60 \end{bmatrix}$$

Step 2: Construct the Normalized Decision Matrix

Similar to the TOPSIS, the normalized decision matrix is calculated as

$$R = \begin{bmatrix} 0.480 & 0.482 & 0.472 \\ 0.552 & 0.468 & 0.558 \\ 0.432 & 0.551 & 0.447 \\ 0.528 & 0.496 & 0.515 \end{bmatrix}$$

Step 3: Construct the Weighted Normalized Matrix

The relative weight matrix W is expressed as

$$W = \begin{bmatrix} 1/3 & 0 & 0 \\ 0 & 1/3 & 0 \\ 0 & 0 & 1/3 \end{bmatrix}$$

By multiplying R with W, the weighted normalized matrix V is estimated as

$$V = \begin{bmatrix} 0.160 & 0.161 & 0.157 \\ 0.184 & 0.156 & 0.186 \\ 0.144 & 0.184 & 0.149 \\ 0.176 & 0.165 & 0.172 \end{bmatrix}$$

Although the first three steps of the ELECTRE method are the same as those of the TOPSIS, the subsequent steps are entirely different for these two methods.

Step 4: Determine the Concordance and Discordance Sets

In order to determine the concordance and discordance sets, each alternative is compared with every other alternative. For an example, when alternative A_1 is compared with alternative A_2, it is observed from the weighted normalized

TABLE 4.2

Concordance and Discordance Sets for Different
Combinations of k and l

K	l	Concordance Set (C_{kl})	Discordance Set (D_{kl})
1	2	{2, 3}	{1}
1	3	{1}	{2, 3}
1	4	{3}	{1, 2}
2	1	{1}	{2, 3}
2	3	{1}	{2, 3}
2	4	{1}	{2, 3}
3	1	{2, 3}	{1}
3	2	{2, 3}	{1}
3	4	{2, 3}	{1}
4	1	{1, 2}	{3}
4	2	{2, 3}	{1}
4	3	{1}	{2, 3}

matrix V that $v_{11} < v_{21}$, $v_{12} > v_{22}$, and $v_{13} < v_{23}$. Since C_1 and C_2 are benefit criteria, and C_3 is the cost criterion, A_1 is preferred over A_2 for criteria C_2 and C_3, whereas A_2 is preferred over A_1 for criterion C_1. Hence, for $k = 1$ and $l = 2$, the concordance set is {2, 3} and the discordance set is {1}. Similarly, for different combinations of k and l, concordance and discordance sets can be calculated, which are tabulated in Table 4.2.

Step 5: Construct the Concordance and Discordance Matrices

The element of kth row and lth column, c_{kl} of the concordance matrix, is obtained by summing the weights associated with the criteria contained in the concordance set. For example, when $k = 1$ and $l = 2$, the criteria 2 and 3 are contained in the concordance set. As each criterion has an equal weight of 1/3, the element c_{12} turns out to be 0.67. Similarly, the other elements of the concordance matrix can be calculated. Hence, the concordance matrix C is obtained as follows:

$$C = \begin{bmatrix} - & 0.67 & 0.33 & 0.33 \\ 0.33 & - & 0.33 & 0.33 \\ 0.67 & 0.67 & - & 0.67 \\ 0.67 & 0.67 & 0.33 & - \end{bmatrix}$$

where the entries of matrix C are not defined when $k = l$.

The element of kth row and lth column, d_{kl} of the discordance matrix, is obtained from Equation 4.10. For an example, when $k = 2$ and $l = 3$, the discordance set is composed of criteria 2 and 3. Thus, the value of d_{23} is calculated as follows:

$$d_{23} = \frac{\max\limits_{j \in D_{23}} |v_{2j} - v_{3j}|}{\max\limits_{j} |v_{2j} - v_{3j}|}$$

$$= \frac{\max\{|0.156 - 0.184|, |0.186 - 0.149|\}}{\max\{|0.184 - 0.144|, |0.156 - 0.184|, |0.186 - 0.149|\}}$$

$$= \frac{\max\{|-0.028|, |0.037|\}}{\max\{|0.04|, |-0.028|, |0.037|\}}$$

$$= \frac{\max\{0.028, 0.037\}}{\max\{0.04, 0.028, 0.037\}}$$

$$= \frac{0.037}{0.04}$$

$$= 0.925$$

Similarly, the other elements of the discordance matrix can be calculated. Thus, the discordance matrix is expressed as follows:

$$D = \begin{bmatrix} - & 0.838 & 1 & 1 \\ 1 & - & 0.925 & 1 \\ 0.697 & 1 & - & 1 \\ 0.895 & 0.559 & 0.716 & - \end{bmatrix}$$

The entries of matrix D are not defined when $k = l$.

Step 6: Determine the Concordance and Discordance Dominance Matrices

The threshold value $\underline{c}$ can be obtained using Equation 4.13 as follows:

$$\underline{c} = \frac{1}{4(4-1)} \sum_{\substack{k=1 \\ k \neq l}}^{4} \sum_{\substack{l=1 \\ l \neq k}}^{4} c_{kl}$$

$$= \frac{(1.67 + 2 + 1 + 1.33)}{4(4-1)}$$

$$= 0.5$$

The elements of the concordance dominance matrix F are determined as follows:

$$f_{kl} = 1, \quad \text{if } c_{kl} \geq 0.5$$

$$f_{kl} = 0, \quad \text{if } c_{kl} < 0.5$$

Hence, concordance dominance matrix can be expressed as

$$F = \begin{bmatrix} - & 1 & 0 & 0 \\ 0 & - & 0 & 0 \\ 1 & 1 & - & 1 \\ 1 & 1 & 0 & - \end{bmatrix}$$

The threshold value $\underline{d}$ is obtained using Equation 4.14 as follows:

$$\underline{d} = \frac{1}{4(4-1)} \sum_{\substack{k=1 \\ k \neq l}}^{4} \sum_{\substack{l=1 \\ l \neq k}}^{4} d_{kl}$$

$$= \frac{(2.592 + 2.397 + 3)}{4(4-1)}$$

$$= 0.886$$

The elements of the concordance dominance matrix F are determined as follows:

$$g_{kl} = 1, \quad \text{if } d_{kl} \geq 0.886$$

$$g_{kl} = 0, \quad \text{if } d_{kl} < 0.886$$

Hence, the concordance dominance matrix can be expressed as

$$G = \begin{bmatrix} - & 0 & 1 & 1 \\ 1 & - & 1 & 1 \\ 0 & 1 & - & 1 \\ 1 & 0 & 0 & - \end{bmatrix}$$

Step 7: Determine the Aggregate Dominant Matrix

The elements of the aggregate dominance matrix E are calculated using Equation 4.15. Thus, the aggregate dominance matrix is obtained as follows:

$$E = \begin{bmatrix} - & 1 & 0 & 0 \\ 0 & - & 0 & 0 \\ 1 & 1 & - & 1 \\ 1 & 1 & 0 & - \end{bmatrix} \cdot \begin{bmatrix} - & 0 & 1 & 1 \\ 1 & - & 1 & 1 \\ 0 & 1 & - & 1 \\ 1 & 0 & 0 & - \end{bmatrix}$$

$$= \begin{bmatrix} - & 0 & 0 & 0 \\ 0 & - & 0 & 0 \\ 0 & 1 & - & 1 \\ 1 & 0 & 0 & - \end{bmatrix}$$

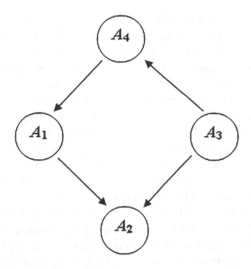

FIGURE 4.1
Design graph of the MCDM problem.

Step 8: Eliminate the Less Favorable Alternatives

From the aggregate dominance matrix, it is observed that except for column 3, all columns contain an element equal to 1; hence, A_3 is considered to be the best alternative. In addition, as $e_{32} = 1$ and $e_{34} = 1$, A_3 is preferred over A_2 and A_4. The value of $e_{41} = 1$ indicates that A_4 is preferred over A_1. Furthermore, by changing the threshold value $\underline{d} = 0.8$, matrix E is modified as follows:

$$
E = \begin{bmatrix}
- & 1 & 0 & 0 \\
0 & - & 0 & 0 \\
0 & 1 & - & 1 \\
1 & 0 & 0 & -
\end{bmatrix}
$$

In the above matrix, $e_{12} = 1$ means that A_1 is preferred to A_2. Thus, the ranks for the alternatives using the ELECTRE method emerge as $A_3 > A_4 > A_1 > A_2$, which is depicted in Figure 4.1. The TOPSIS for the same problem as discussed in the preceding chapter yields the ranking of the alternatives as $A_4 > A_3 > A_1 > A_2$.

The MATLAB® coding for the aforesaid MCDM problem using the ELECTRE method is given in Section 4.5.

4.4 Application of ELECTRE Method in Textiles

There are some reported works on the application of the ELECTRE method in the field of textiles. Ghosh et al. (2012) applied the ELECTRE method in cotton fiber grading. The weights of different decision criteria and subcriteria were evaluated by the analytic hierarchy process (AHP) and then the ELECTRE method was employed to elicit the final ranking of cotton fibers. Twenty-five types of cottons were used as the alternatives. Three properties of cotton fibers, viz., tensile, length, and fineness, were used as the criteria. The tensile properties were divided into two subcriteria: bundle strength (FS) and bundle elongation (FE). The upper-half mean length (UHML), length uniformity index (UI), and short fiber content (SFC) were the subcriteria for the length properties. The fineness property was exclusively represented by the micronaire value (MIC). The ranking of cotton fibers attained by the ELECTRE method shows significant agreement with the ranking based on the strength of the yarn. The performance of the ELECTRE method was found to be better than the TOPSIS for grading cotton fibers.

Kaplan et al. (2006) established that the application of the ELECTRE III method can provide valuable assistance in reaching acceptable solutions in order to select the appropriate rotor navel for the yarns intended for denim fabrics. Ten navels having different technical properties were ranked according to their performances in yarn quality by using the ELECTRE III method.

Özdemır (2017) used fuzzy AHP-ELECTRE methods in the supplier selection process. A large firm that operates in the textile sector has been selected, and important criteria have been identified with strategic decision makers. After that, current suppliers were evaluated, and the most suitable suppliers for the company were identified.

4.4.1 Selection of Bulletproof Body Armors Using ELECTRE Method

Majumdar et al. (2011) used the AHP of MCDM for the selection of bulletproof body armor based on three decision criteria, viz., impact resistance, comfort score, and cost. The impact resistance of body armor is characterized by the V_{50} speed at which the bullet has equal probability to pierce the vest or to be stopped by the vest. Comfort score has been taken as an overall index representing the flexibility, thermal resistance, and moisture vapor transmission of the body armor. Higher V_{50} speed is a desirable or beneficial criterion and so is the comfort score. However, price of the vest is a negative or cost criterion, and a lower value is desirable. The features of the bulletproof body armors are shown in Table 4.3.

In this chapter, the ELECTRE method of MCDM has been demonstrated for the selection of body armor using the data of Majumdar et al. (2011) as depicted in Table 4.3. Figure 4.2 shows the hierarchical structure of this MCDM problem. The relative weights of the three criteria were determined using the AHP method as discussed later.

Table 4.4 shows the pair-wise comparison matrix of three decision criteria based on the perception of the decision maker. Here numerical scores have been given as per Saaty's nine-point scale as described in Chapter 2 (Saaty, 1980). Impact resistance has moderate dominance over the comfort, and

TABLE 4.3

Features of Bulletproof Body Armors

Alternatives	Impact Resistance (V_{50}, m/s)	Comfort Score	Cost (INR)
A_1	450	1,000	40,000
A_2	500	1,500	50,000
A_3	475	800	45,000
A_4	400	2,000	45,000

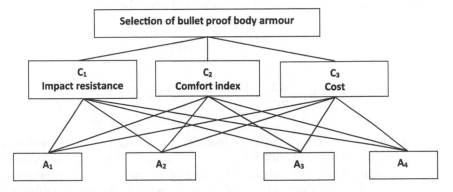

FIGURE 4.2

The hierarchical structure of the MCDM problem.

TABLE 4.4

Pair-Wise Comparison Matrix

Criteria	Impact Resistance	Comfort Score	Cost	Geometric Mean (GM)	Normalized GM or Relative Weights
Impact resistance	1	3	5	2.46	0.64
Comfort score	1/3	1	3	1	0.26
Cost	1/5	1/3	1	0.41	0.10

comfort has moderate dominance over the cost. Cost has the least influence on the decision as high-impact resistance and greater comfort are imperative for body armors. After calculating the normalized geometric mean of rows, it has been found that the weights of impact resistance, comfort score, and cost are 0.64, 0.26, and 0.10, respectively. For the measurement of consistency of judgment, the original pair-wise comparison matrix was multiplied by the priority vector or weight vector to get the product as shown:

$$\begin{bmatrix} 1 & 3 & 5 \\ 1/3 & 1 & 3 \\ 1/5 & 1/3 & 1 \end{bmatrix} \times \begin{bmatrix} 0.64 \\ 0.26 \\ 0.10 \end{bmatrix} = \begin{bmatrix} 1.920 \\ 0.773 \\ 0.315 \end{bmatrix}$$

The principal eigenvector (λ_{max}) of the original pair-wise comparison matrix is calculated as follows:

$$\lambda_{max} = \frac{\left(\dfrac{1.920}{0.64} + \dfrac{0.773}{0.26} + \dfrac{0.315}{0.10}\right)}{3} = 3.041$$

To check the consistency in pair-wise comparison, the consistency index (CI) and the consistency ratio (CR) are calculated from the following formulas:

$$CI = \frac{3.041 - 3}{3 - 1} = 0.0205$$

$$CR = \frac{CI}{RCI} = \frac{0.0205}{0.58} = 0.0353 < 0.1$$

where RCI is the random consistency index and for three criteria, its value is 0.58. As the calculated value of CR is less than 0.1, the judgment is considered to be consistent. After determining the weights of the decision criteria, the steps of the ELECTRE method are followed as explained earlier. Tables 4.5 and 4.6 show the normalized and weighted normalized features of the four body armors, respectively.

For various combinations of k and l, concordance and discordance sets are illustrated in Table 4.7.

With regard to the weights associated with each criteria, the concordance matrix is represented as follows:

$$C = \begin{bmatrix} - & 0.1 & 0.36 & 0.74 \\ 0.9 & - & 0.9 & 0.64 \\ 0.64 & 0.1 & - & 0.74 \\ 0.26 & 0.36 & 0.36 & - \end{bmatrix}$$

TABLE 4.5

Normalized Features of Body Armors

Alternatives	Impact Resistance	Comfort Score	Cost
A_1	0.492	0.356	0.443
A_2	0.546	0.534	0.554
A_3	0.519	0.285	0.499
A_4	0.437	0.712	0.499

TABLE 4.6

Weighted Normalized Features of Body Armors

Alternatives	Impact Resistance	Comfort Score	Cost
A_1	0.315	0.093	0.044
A_2	0.350	0.139	0.055
A_3	0.332	0.074	0.050
A_4	0.280	0.185	0.050

TABLE 4.7

Concordance and Discordance Sets

K	l	Concordance Set (C_{kl})	Discordance Set (D_{kl})
1	2	{3}	{1, 2}
1	3	{2, 3}	{1}
1	4	{1, 3}	{2}
2	1	{1, 2}	{3}
2	3	{1, 2}	{3}
2	4	{1}	{2, 3}
3	1	{1}	{2, 3}
3	2	{3}	{1, 2}
3	4	{1, 3}	{2}
4	1	{2}	{1, 3}
4	2	{2, 3}	{1}
4	3	{2, 3}	{1}

The elements of the discordance matrix can be calculated from Equation 4.10, which yields the discordance matrix as follows:

$$D = \begin{bmatrix} - & 1 & 0.944 & 1 \\ 0.239 & - & 0.085 & 0.662 \\ 1 & 1 & - & 1 \\ 0.378 & 1 & 0.472 & - \end{bmatrix}$$

The threshold value from the concordance matrix is estimated as $\underline{c} = 0.508$, which leads to the concordance dominance matrix as follows:

$$F = \begin{bmatrix} - & 0 & 0 & 1 \\ 1 & - & 1 & 1 \\ 1 & 0 & - & 1 \\ 0 & 0 & 0 & - \end{bmatrix}$$

The discordance matrix generates threshold value $\underline{d} = 0.732$; thus, the discordance dominance matrix is represented as follows:

$$G = \begin{bmatrix} - & 1 & 1 & 1 \\ 0 & - & 0 & 0 \\ 1 & 1 & - & 1 \\ 0 & 1 & 0 & - \end{bmatrix}$$

The aggregate dominance matrix E is constructed from the elements of the matrices F and G using Equation 4.15. Therefore, the aggregate dominance matrix is obtained as follows:

$$E = \begin{bmatrix} - & 0 & 0 & 1 \\ 0 & - & 0 & 0 \\ 1 & 0 & - & 1 \\ 0 & 0 & 0 & - \end{bmatrix}$$

From the aggregate dominance matrix, it is noted that columns 1 and 4 contain at least an element equal to 1; hence, alternatives A_1 and A_4 are outperformed by alternatives A_2 and A_3. Moreover, the elements e_{14}, e_{31}, and e_{34} are equal to 1; thus, A_1 is preferred over A_4, and A_3 is preferred over A_1 and A_4. However, it is inconclusive to find out the best alternatives between A_2 and A_3. This may be ascribed to the fact that the ELECTRE method primarily yields a system of binary outranking relations between the alternatives, which basically eliminates the less favorable alternatives; however, sometimes it may not be able to identify the most favorable alternative.

4.5 MATLAB® Coding

MATLAB® Coding for the ELECTRE Method

```
clc
clear all
close all
%Decision matrix(a)
a = [100    35    55
      115   34    65
       90   40    52
      110   36    60];

%cost and benefit criteria
cc = 3;
bc = 1:2;
```

```
[s1,s2] = size(a);
%Step1: Calculate normalized decision matrix (X)
c = a.^2;
j = 1 : s2;
s = sqrt(sum(c(:,j)));
for j = 1 : s2;
X(:,j) = (a(:,j))/s(j);
end
%Step 2: Calculate weighted normalized decision matrix (y)
w = [1/3 1/3 1/3];
W = diag(w);
y = X*W
%Name the alternatives K and L for binary comparison
for K = 1 : s1
    for L = 1 : s1
        [K L];
        if K = = L
            %Step 3:Determination of concordance set and
concordance matrix
            C_kl = 0;
        else
            C_kl = 0;
            %For profit criteria
            for j = bc
                d(j) = y(K,j)-y(L,j);
                if d(j) > = 0
                    Concordance_set = j;
                    C_kl = C_kl + w(j);
                end
            end
            %For cost criteria
            for j = cc
                d(j) = y(K,j) - y(L,j);
                if d(j) < = 0
                    Concordance_set = j;
                    C_kl = C_kl + w(j);
                end
            end
        end
        C(K,L) = C_kl;
    end
end
Concordance_matrix = C
for K = 1 : s1
    for L = 1 : s1
        [K L];
        if K = = L
            D_kl = 0;
        else
```

```
            %Step 4: Determination of discordance set and
discordance matrix
            d = y(K,:) - y(L,:);
            A = abs(d);
            den = max(A);
            %For profit criteria
            for j=bc
                d(j) = y(K,j) - y(L,j);
                if d(j) < 0
                    num(j) = abs(d(j));
                    Discordance_set = j;
                    D_kl(j) = num(j)/den;
                else
                    D_kl(j) = 0;
                end
            end
            %For cost criteria
            for j = cc
                d(j) = y(K,j) - y(L,j);
                if d(j) > 0
                    Discordance_set = j;
                    num(j) = abs(d(j));
                    D_kl(j) = num(j)/den;
                else
                    D_kl(j) = 0;
                end
            end
        end
        D(K,L) = max(D_kl);
    end
end
Discordance_matrix = D
[m,n] = size(C);
c_bar = (sum(sum(C)))/(m*(m-1))
d_bar = (sum(sum(D)))/(m*(m-1))
for i = 1 : m
    for j = 1 : n
        if C(i,j) > = c_bar
        F(i,j) = 1;
        else
        F(i,j) = 0;
        end
    end
end
F
for i=1:m
    for j=1:n
        if D(i,j) > = d_bar
        G(i,j) = 1;
        else
```

```
        G(i,j) = 0;
        end
    end
end
G
 E=F.*G
```

4.6 Summary

This chapter presents different steps of the ELECTRE method. A worked out problem in step-by-step detail has been discussed along with the MATLAB® coding. A survey of literature on the application of the ELECTRE method in textiles shows that only limited works have been carried out so far. This chapter also includes a detailed study on the selection of bulletproof body armors using the ELECTRE method.

References

Abedi, M., Torabi, S. A., Norouzi, G. H. and Hamzeh, M. 2012. ELECTRE III: A knowledge driven method for integration of geophysical data with geological and geochemical data in mineral prospectivity mapping. *Journal of Applied Geophysics*, 87, 9–18.

Benayoun, R., Roy, B. and Sussman, N. 1966. *Manual de Reference du Program ELECTRE*, Note de Synthese et Formation, Direction Scientifque, SEMA, No. 25, Paris, France.

Damaskos, X. and Kalfakakou, G. 2005. Application of ELECTRE III and DEA methods in the BRP of a bank branch network. *Yugoslav Journal of Operations Research*, 15(2), 259–276.

Ghosh, A., Majumdar, A. and Alam, S. 2012. Selection of raw materials in textile spinning industry using ELECTRE. *Industrial Engineering Journal*, 5(6), 6–15.

Kaplan, S., Araz, C. and Göktepe, Ö. 2006. A multicriteria decision aid approach on navel selection problem for rotor spinning. *Textile Research Journal*, 76(12), 896–904.

Majumdar, A., Singh, S. P. and Ghosh, A. 2011. Modelling, optimization and decision making techniques in designing of functional clothing. *Indian Journal of Fibre and Textile Research*, 36, 398–409.

Özdemır, Y. S. 2017. Supplier selection by using fuzzy AHP-ELECTRE and an application in textile company. *Proceedings of the IEEE Conference on Electric Electronics, Computer Science, Biomedical Engineerings' Meeting (EBBT)*, Istanbul, Turkey.

Papadopoulos, A. and Karagiannidis, A. 2008. Application of the multi-criteria analysis method ELECTRE III for the optimisation of decentralised energy systems. *Omega*, 36, 766–776.

Roy, B. 1991. The outranking approach and the foundations of ELECTRE methods. *Theory and Decision*, 31, 49–73.

Saaty, T. L. 1980. *The Analytic Hierarchy Process*, McGraw-Hill, New York.

Sánchez-Lozano, J., Henggeler Antunes, C., García-Cascales, M. and Dias, L. 2014. GIS based photovoltaic solar farms site selection using ELECTRE-TRI: Evaluating the case for Torre Pacheco, Murcia, southeast of Spain. *Renewable Energy*, 66, 478–494.

Triantaphyllou, E. 2000. *Multi-Criteria Decision Making Methods: A Comparative Study*, Kluwer Academic Publishers, Boston.

5

Graph Theory and Matrix
Approach of Decision-Making

5.1 Introduction

The graph theory and matrix approach (GTMA) is a method of multicriteria decision-making (MCDM) that is derived from combinatorial mathematics. The GTMA method uses a graph that represents the selection criteria and their interdependencies in terms of nodes and edges, whereas the matrix translates the graph into mathematical form. The rank of the alternatives is determined using a permanent function. The GTMA method was first adopted by Gandhi and Agrawal (1992, 1994) for assessing the failure mode and effects analysis of mechanical systems. Rao and Gandhi (2001, 2002) applied GTMA in the field of machining and manufacturing to evaluate the machinability index. Rao (2007) presented the methodology of GTMA in detail. The GTMA method has become an increasingly significant MCDM tool in the fields of engineering and technology.

5.2 Graph Theory and Matrix Approach

A graph is composed of two sets of objects, commonly represented as $G = (N, A)$, where the set $N = \{N_1, N_2, \ldots, N_i, \ldots, N_M\}$, whose elements are termed as *nodes*, and another set $A = \{a_{12}, a_{21}, \ldots, a_{ij}, a_{ji}, \ldots\}$, elements of which are termed as *edges* (Deo, 2000). The graph is generally represented by means of a diagram as shown in Figure 5.1, in which nodes are denoted by circles and edges are represented by curve segments joining the nodes. For the sake of simplicity, only four nodes are depicted in Figure 5.1. A node N_i indicates the ith criterion, and an edge is a representation of relative importance between a pair of criteria. A directed edge (a_{ij}) from ith node to jth node implies the relative importance of the ith criterion over the jth criterion. Similarly, the

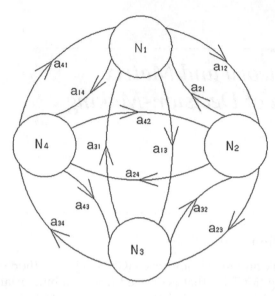

FIGURE 5.1
Representation of a graph.

relative importance of the jth criterion over the ith criterion is represented by the directed edge a_{ji}.

A matrix representation of the selection criteria gives one-to-one representation. For M number of criteria, an $M \times M$ matrix, which is termed as *relative importance matrix of criteria* (E), is constructed, where an element a_{ij} is the relative importance of the ith criterion over the jth one. The pair-wise comparison matrix of criteria is expressed as follows:

$$E = \begin{bmatrix} - & a_{12} \cdots a_{1j} \cdots & a_{1M} \\ a_{21} & - \cdots a_{2j} \cdots & a_{2M} \\ \vdots & \cdots & \vdots \\ a_{j1} & a_{j2} \cdots - \cdots & a_{jM} \\ \vdots & \cdots & \vdots \\ a_{M1} & a_{M2} \cdots a_{Mj} \cdots & - \end{bmatrix} \tag{5.1}$$

The principle of the GTMA of MCDM is discussed in the following steps.

Step 1: Construct the Decision Matrix

This step constructs the decision matrix D of K alternatives and M criteria with the information available regarding the MCDM problem as follows:

$$D = \begin{bmatrix} x_{11} & x_{12} & \cdots & x_{1j} & \cdots & x_{1M} \\ x_{21} & x_{22} & \cdots & x_{2j} & \cdots & x_{2M} \\ \vdots & \cdots & \cdots & \cdots & \cdots & \vdots \\ x_{p1} & x_{p2} & \cdots & x_{pj} & \cdots & x_{pM} \\ \vdots & \cdots & \cdots & \cdots & \cdots & \vdots \\ x_{K1} & x_{K2} & \cdots & x_{Kj} & \cdots & x_{KM} \end{bmatrix}$$ (5.2)

where x_{pj} denotes the performance measure of the pth alternative in terms of the jth criterion.

Step 2: Construct the Normalized Decision Matrix

This step transforms the entries of the decision matrix into dimensionless comparable entries. For a benefit criterion j, the normalized value of the pth alternative is estimated as follows:

$$r_{pj} = \frac{x_{pj}}{x_{uj}}$$ (5.3)

where x_{uj} is the maximum value of the jth criteria among all of the alternatives. In case of a cost criterion j, the normalized value of the pth alternative can be estimated as

$$r_{pj} = \frac{x_{lj}}{x_{pj}}$$ (5.4)

where x_{lj} is the lowest value of the jth criteria among all of the alternatives. Thus, the normalized decision matrix R is expressed as

$$R = \begin{bmatrix} r_{11} & r_{12} & \cdots & r_{1j} & \cdots & r_{1M} \\ r_{21} & r_{22} & \cdots & r_{2j} & \cdots & r_{2M} \\ \vdots & \cdots & \cdots & \cdots & \cdots & \vdots \\ r_{p1} & r_{p2} & \cdots & r_{pj} & \cdots & r_{pM} \\ \vdots & \cdots & \cdots & \cdots & \cdots & \vdots \\ r_{K1} & r_{K2} & \cdots & r_{Kj} & \cdots & r_{KM} \end{bmatrix}$$ (5.5)

Step 3: Construct the Relative Importance Matrix of Criteria

The relative importance matrix of criteria (E) is represented in Equation 5.1. The relative importance between two criteria (a_{ij}) for a given MCDM problem is distributed on the scale 0–1. If a_{ij} denotes the relative importance of the ith criterion over the jth criterion, then the relative importance of the jth criterion over the ith criterion, a_{ji}, is estimated as follows:

$$a_{ji} = 1 - a_{ij} \tag{5.6}$$

In case of three criteria, the relative importance is expressed in three classes as shown in Table 5.1. Tables 5.2 and 5.3 show the relative importance values in case of five and six criteria, respectively.

TABLE 5.1

Relative Importance Value on a Three-Point Scale

	Relative Importance	
Class Description	a_{ij}	$a_{ji} = 1 - a_{ij}$
One criterion is less important than the other.	0.17	0.83
Two criteria are equally important.	0.50	0.50
One criterion is more important than the other.	0.83	0.17

TABLE 5.2

Relative Importance Value on a Five-Point Scale

	Relative Importance	
Class Description	a_{ij}	$a_{ji} = 1 - a_{ij}$
One criterion is very less important than the other.	0.11	0.89
One criterion is less important than the other.	0.30	0.70
Two criteria are equally important.	0.50	0.50
One criterion is more important than the other.	0.70	0.30
One criterion is much more important than the other.	0.89	0.11

TABLE 5.3

Relative Importance Value on a Six-Point Scale

	Relative Importance	
Class Description	a_{ij}	$a_{ji} = 1 - a_{ij}$
Two criteria are equally important.	0.5	0.5
One criterion is slightly more important than the other.	0.6	0.4
One criterion is strongly more important than the other.	0.7	0.3
One criterion is very strongly more important than the other.	0.8	0.2
One criterion is extremely more important than the other.	0.9	0.1
One criterion is exceptionally more important than the other.	1.0	0.0

Source: Rao, R. V. 2007. *Decision Making in the Manufacturing Environment: Using Graph Theory and Fuzzy Multiple Attribute Decision Making Methods*, Springer, London.

Step 4: Determine the Relative Importance Matrix for Each Alternative

In this step, the diagonal of the matrix E (Equation 5.1) is filled with the normalized value of each alternative. Thus, for the pth alternative, the relative importance matrix H_p is written as follows:

$$H_p = \begin{vmatrix} r_{p1} & a_{12}\cdots a_{1j}\cdots & a_{1M} \\ a_{21} & r_{p2}\cdots a_{2j}\cdots & a_{2M} \\ \vdots & \cdots & \vdots \\ a_{j1} & a_{j2}\cdots r_{pj} \cdots & a_{jM} \\ \vdots & \cdots & \vdots \\ a_{M1} & a_{M2}\cdots a_{Mj}\cdots & r_{pM} \end{vmatrix} \tag{5.7}$$

where r_{pj} is the normalized value of the jth criterion for the pth alternative. It is noted that for each alternative, there is a separate relative importance matrix.

Step 5: Determine the Permanent of the Relative Importance Matrix for Each Alternative

In this step, the permanent of the relative importance matrix for each alternative is calculated. The permanent of an $M \times M$ square matrix H_p is determined as follows:

$$\text{per}(H_p) = \sum_{\sigma \in S_M} \prod_{j=1}^{M} a_{j,\sigma(j)} \tag{5.8}$$

where the sum is extended over all elements of the symmetric group S_M, that is, over all permutations of the numbers 1, 2, 3, ... , M. If the sign of a permutation $\sigma \in S_M$ is expressed by

$$\text{sgn}(\sigma) = \begin{cases} 1, & \text{for } \sigma \text{ is even} \\ -1, & \text{for } \sigma \text{ is odd} \end{cases},$$

then the determinant of the matrix H_p is defined as follows:

$$\det(H_p) = \sum_{\sigma \in S_M} \text{sgn}(\sigma) \prod_{j=1}^{M} a_{j,\sigma(j)} \tag{5.9}$$

For example, in case of a general 3×3 matrix,

$$A = \begin{bmatrix} a & b & c \\ d & e & f \\ g & h & i \end{bmatrix}$$

$$\det(A) = aei - afh - bdi + bfg + cdh - ceg$$

$$\mathrm{per}(A) = aei + afh + bdi + bfg + cdh + ceg$$

Step 6: Rank the Preference Order

All of the alternatives are now arranged in descending order according to the values of $\mathrm{per}(H_p)$. Therefore, the best and worst alternatives have the highest and lowest values of $\mathrm{per}(H_p)$.

5.3 Step-by-Step Working Principles of Graph Theory and Matrix Approach

The same example of the MCDM problem as discussed in Chapter 3, Section 3.3 has been considered here to explain the working principles of the GTMA in step-by-step fashion.

Step 1: Construct the Decision Matrix

The decision matrix is represented as follows:

$$D = \begin{bmatrix} 100 & 35 & 55 \\ 115 & 34 & 65 \\ 90 & 40 & 52 \\ 110 & 36 & 60 \end{bmatrix}$$

Step 2: Construct the Normalized Decision Matrix

In this MCDM problem, the first two columns of matrix D show benefit criteria, and the last column shows the cost criteria. Hence, using Equations 5.3 and 5.4, we obtain the normalized decision matrix as follows:

$$R = \begin{bmatrix} 0.8696 & 0.875 & 0.9455 \\ 1 & 0.85 & 0.80 \\ 0.7826 & 1 & 1 \\ 0.9565 & 0.90 & 0.8667 \end{bmatrix}$$

Step 3: Construct the Relative Importance Matrix of Criteria

In this example, we considered that each criterion has equal importance. Hence, the relative importance matrix of criteria can be expressed as follows:

$$E = \begin{bmatrix} - & 0.5 & 0.5 \\ 0.5 & - & 0.5 \\ 0.5 & 0.5 & - \end{bmatrix}$$

Step 4: Determine the Relative Importance Matrix for Each Alternative

In this step, the diagonal of the matrix E is filled with the normalized value of each alternative. The relative importance matrices for one to four alternatives are expressed as follows:

$$H_1 = \begin{bmatrix} 0.8696 & 0.5 & 0.5 \\ 0.5 & 0.875 & 0.5 \\ 0.5 & 0.5 & 0.9455 \end{bmatrix}$$

$$H_2 = \begin{bmatrix} 1 & 0.5 & 0.5 \\ 0.5 & 0.85 & 0.5 \\ 0.5 & 0.5 & 0.8 \end{bmatrix}$$

$$H_3 = \begin{bmatrix} 0.7826 & 0.5 & 0.5 \\ 0.5 & 1 & 0.5 \\ 0.5 & 0.5 & 1 \end{bmatrix}$$

$$H_4 = \begin{bmatrix} 0.9565 & 0.5 & 0.5 \\ 0.5 & 0.90 & 0.5 \\ 0.5 & 0.5 & 0.8667 \end{bmatrix}$$

Step 5: Determine the Permanent of the Relative Importance Matrix for Each Alternative

The permanent values of the relative importance matrices for each alternative are estimated as follows:

$\text{per}(H_1) = 1.642$
$\text{per}(H_2) = 1.5925$
$\text{per}(H_3) = 1.7283$
$\text{per}(H_4) = 1.6769$

Step 6: Rank the Preference Order

From the permanent values, alternative 3 is considered to be the best choice followed by alternatives 4, 1, and 2.

The MATLAB® coding for the aforesaid MCDM problem using the GTMA is given in Section 5.5.

5.4 Application of Graph Theory and Matrix Approach of Decision-Making in Textiles

There are not many published reports on the application of the GTMA in the field of textiles. In the following example, we have applied it for the selection of cotton fibers.

5.4.1 Cotton Fiber Selection Using Graph Theory and Matrix Approach

The GTMA has been used to determine the quality value of cotton fibers with a goal to achieve maximum yarn strength. Five types of cottons are used for this purpose. Six properties of cotton fibers, viz., strength, elongation, upper-half mean length (UHML), length uniformity index (UI), fineness, and short fiber content (SFC), have been used as the criteria for this specific problem. The fiber samples were collected from five types of cotton populations by using a random sampling technique. Each sample was conditioned at standard atmospheric conditions for 24 hours before the testing. Fiber strength, fineness, UHML, and UI were tested in the length/strength module of the Uster high-volume instrument (HVI 900). The fineness was measured in the micronaire module of HVI 900. The SFC was measured by the Uster Advanced Fibre Information System (AFIS). Higher values of strength, elongation, UHML, and UI are desirable, so these are considered as the benefit criteria. However, for fineness and SFC, lower values are desirable; hence, they are treated as cost criteria. The properties of the five types of cotton fibers are shown in Table 5.4. The hierarchical structure of the cotton selection problem is depicted in Figure 5.2.

TABLE 5.4

Properties of Five Types of Cotton Fibers

Alternatives	Strength, cN/tex (C_1)	Elongation, % (C_2)	UHML, inch (C_3)	UI, % (C_4)	Fineness, μg/inch (C_5)	SFC, % (C_6)
A_1	28.7	6.5	27.69	81.0	4.4	13.8
A_2	31.7	6.3	26.16	80.6	3.7	8.9
A_3	29.1	6.9	26.67	83.2	4.6	5.6
A_4	30.8	6.4	25.65	81.7	3.7	6.8
A_5	26.7	6.9	26.42	82.6	4.8	7.5

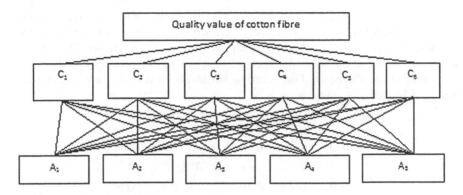

FIGURE 5.2
Hierarchical structure of the cotton selection problem.

The decision matrix of the cotton fiber selection problem is expressed as follows:

$$D = \begin{bmatrix} 28.7 & 6.5 & 27.69 & 81.0 & 4.4 & 13.8 \\ 31.7 & 6.3 & 26.16 & 80.6 & 3.7 & 8.9 \\ 29.1 & 6.9 & 26.67 & 83.2 & 4.6 & 5.6 \\ 30.8 & 6.4 & 25.65 & 81.7 & 3.7 & 6.8 \\ 26.7 & 6.9 & 26.42 & 82.6 & 4.8 & 7.5 \end{bmatrix}$$

In the next step, the normalized decision matrix is constructed. As the first four criteria of the decision matrix are benefit types, the maximum values become one after normalization. The last two criteria of the decision matrix are cost types; hence, the minimum values become one after normalization. The normalized decision matrix is represented as follows:

$$R = \begin{bmatrix} 0.91 & 0.94 & 1.00 & 0.97 & 0.84 & 0.41 \\ 1.00 & 0.91 & 0.94 & 0.97 & 1.00 & 0.63 \\ 0.92 & 1.00 & 0.96 & 1.00 & 0.80 & 1.00 \\ 0.97 & 0.93 & 0.93 & 0.98 & 1.00 & 0.82 \\ 0.84 & 1.00 & 0.95 & 0.99 & 0.77 & 0.75 \end{bmatrix}$$

Then, we carry out pair-wise comparison among the six criteria according to the relative importance values on a six-point scale as shown in Table 5.3. The UHML is considered to be the most important of the criteria. The UHML is assessed to have slightly more importance than strength; strongly more importance than UI, fineness, and SFC; and extreme importance over elongation. Strength is regarded as very strongly more important than elongation and fineness but slightly more important than UI and SFC. The UI is considered as strongly more important than elongation and slightly more important than fineness. The SFC and UI are treated as equally important. Fineness is rated as slightly more important than elongation. The SFC is evaluated as strongly more important than elongation and slightly more important than fineness. The relative importance matrix of criteria thus prepared is given as follows:

$$
E = \begin{bmatrix}
- & 0.80 & 0.40 & 0.60 & 0.80 & 0.60 \\
0.20 & - & 0.10 & 0.30 & 0.40 & 0.30 \\
0.60 & 0.90 & - & 0.70 & 0.70 & 0.70 \\
0.40 & 0.70 & 0.30 & - & 0.60 & 0.50 \\
0.20 & 0.60 & 0.30 & 0.40 & - & 0.40 \\
0.40 & 0.70 & 0.30 & 0.50 & 0.60 & -
\end{bmatrix}
$$

Subsequently, the diagonal of the matrix E is filled with the normalized values of each alternative. The following matrices H_1 to H_5 are the relative importance matrices for alternatives A_1 to A_5, respectively:

$$
H_1 = \begin{bmatrix}
0.91 & 0.80 & 0.40 & 0.60 & 0.80 & 0.60 \\
0.20 & 0.94 & 0.10 & 0.30 & 0.40 & 0.30 \\
0.60 & 0.90 & 1.00 & 0.70 & 0.70 & 0.70 \\
0.40 & 0.70 & 0.30 & 0.97 & 0.60 & 0.50 \\
0.20 & 0.60 & 0.30 & 0.40 & 0.84 & 0.40 \\
0.40 & 0.70 & 0.30 & 0.50 & 0.60 & 0.41
\end{bmatrix}
$$

$$
H_2 = \begin{bmatrix}
1.00 & 0.80 & 0.40 & 0.60 & 0.80 & 0.60 \\
0.20 & 0.91 & 0.10 & 0.30 & 0.40 & 0.30 \\
0.60 & 0.90 & 0.94 & 0.70 & 0.70 & 0.70 \\
0.40 & 0.70 & 0.30 & 0.97 & 0.60 & 0.50 \\
0.20 & 0.60 & 0.30 & 0.40 & 1.00 & 0.40 \\
0.40 & 0.70 & 0.30 & 0.50 & 0.60 & 0.63
\end{bmatrix}
$$

$$
H_3 = \begin{bmatrix}
0.92 & 0.80 & 0.40 & 0.60 & 0.80 & 0.60 \\
0.20 & 1.00 & 0.10 & 0.30 & 0.40 & 0.30 \\
0.60 & 0.90 & 0.96 & 0.70 & 0.70 & 0.70 \\
0.40 & 0.70 & 0.30 & 1.00 & 0.60 & 0.50 \\
0.20 & 0.60 & 0.30 & 0.40 & 0.80 & 0.40 \\
0.40 & 0.70 & 0.30 & 0.50 & 0.60 & 1.00
\end{bmatrix}
$$

$$H_4 = \begin{bmatrix} 0.97 & 0.80 & 0.40 & 0.60 & 0.80 & 0.60 \\ 0.20 & 0.93 & 0.10 & 0.30 & 0.40 & 0.30 \\ 0.60 & 0.90 & 0.93 & 0.70 & 0.70 & 0.70 \\ 0.40 & 0.70 & 0.30 & 0.98 & 0.60 & 0.50 \\ 0.20 & 0.60 & 0.30 & 0.40 & 1.00 & 0.40 \\ 0.40 & 0.70 & 0.30 & 0.50 & 0.60 & 0.82 \end{bmatrix}$$

$$H_5 = \begin{bmatrix} 0.84 & 0.80 & 0.40 & 0.60 & 0.80 & 0.60 \\ 0.20 & 1.00 & 0.10 & 0.30 & 0.40 & 0.30 \\ 0.60 & 0.90 & 0.95 & 0.70 & 0.70 & 0.70 \\ 0.40 & 0.70 & 0.30 & 0.99 & 0.60 & 0.50 \\ 0.20 & 0.60 & 0.30 & 0.40 & 0.77 & 0.40 \\ 0.40 & 0.70 & 0.30 & 0.50 & 0.60 & 0.75 \end{bmatrix}$$

The permanent values of the relative importance matrices for alternatives A_1 to A_5 are calculated as follows:

$per(H_1) = 15.33$
$per(H_2) = 17.69$
$per(H_3) = 19.28$
$per(H_4) = 18.87$
$per(H_5) = 16.86$

From the permanent values, A_3 is selected as the best alternative followed by the alternatives A_4, A_2, A_5, and A_1.

5.5 MATLAB® Coding

MATLAB® Coding for Graph Theory and Matrix Approach

```
function r = permanent(z)
[m, n] = size(z);
   if n == 1
      r = z;
   else
      for j = 2 : n
          w = ones(1, j);
          for i = 1 : j
```

```
                R = z( [1: j-1],[1: i-1 i+1: j] );
                w (i) = permanent(R);
            end
            r = z(j,1: j) * w';
        end
    end
clc


clear all
close all
format short g
x = [100        35        55
        115        34        65
         90        40        52
        110        36        60];
N = [x(:,1)./max(x(:,1)) x(:,2)./max(x(:,2))
min(x(:,3))./x(:,3)];
E = [0          0.5       0.5
        0.5     0         0.5
        0.5     0.5       0];
[r, c] = size(N);
for i = 1: r
    y = diag(N(i,:));
    z = E + y;
s(i) = permanent(z);
end
score = s'
```

5.6 Summary

This chapter presents an overview of the GTMA. The different steps of this method are explained, and a worked-out problem and its MATLAB® coding are also provided. This chapter also includes discussion of the selection of cotton fibers using the GTMA.

References

Deo, N. 2000. *Graph Theory with Applications to Engineering and Computer Science*, Prentice Hall, New Delhi, India.
Gandhi, O. P. and Agrawal, V. P. 1992. FMEA—A digraph and matrix approach. *Reliability Engineering and System Safety*, 35, 147–158.

Gandhi, O. P. and Agrawal, V. P. 1994. A digraph approach to system wear evaluation and analysis. *Journal of Tribology*, 116, 268–274.

Rao, R. V. 2007. *Decision Making in the Manufacturing Environment: Using Graph Theory and Fuzzy Multiple Attribute Decision Making Methods*, Springer, London.

Rao, R. V. and Gandhi, O. P. 2001. Digraph and matrix method for selection, identification and comparison of metal cutting fluids. *Proceedings of the Institute of Mechanical Engineers, Part J: Journal of Engineering Tribology*, 215, 25–33.

Rao, R. V. and Gandhi, O. P. 2002. Digraph and matrix methods for machinability evaluation of work materials. *International Journal of Machine Tools and Manufacture*, 42, 321–330.

6

Linear Programming

6.1 Introduction

In engineering design, manufacturing, and business decision-making, it is often necessary to find one or more feasible solutions that will either maximize or minimize one or more objectives. The scientific and quantitative approach that produces the best results in such objective(s) by selecting a suitable combination of decisive variables is known as *optimization*. Like in other engineering design, textile engineering also requires optimization, which is necessary to maximize production, utilization of manpower and machine, and realization of raw material, or to minimize waste and cost, or others. When an optimization problem involves only one objective, the task of finding the optimal solution(s) of such a problem is known as *single-objective optimization*. In other words, a single-objective optimization problem requires an objective function to be minimized or maximized and fulfills other conditions termed *constraints*. Now, if both the *objective* and *constraints* of an optimization problem are linear, then such an optimization problem is classified as a *linear optimization problem*. If either the objective function or the constraints have nonlinearity, then such optimization problems belong to nonlinear optimization. Various algorithms exist to solve these optimization problems. Linear programming (LP) is one of the most popular, versatile, commonly used, and simplest classical optimization techniques to provide an optimal solution for a linear function of decisive variables satisfying linear constraints (Murthy 1983; Sharma 2010; Srinivasan 2014; Taha 1989). This chapter presents a detailed application of methods of LP to deal with linear single-objective optimization problems in textile engineering.

6.2 Linear Programming Problem Formulation

In order to solve an LP problem, it is initially essential to formulate the mathematical model of LP. Consequently, the LP can be solved using various methods. To formulate an LP problem, the unknown decisive variables are identified first. Then the objective of the problem, along with the constraint(s) are identified. The objective is put as a linear function of the decisive variables. The constraints are expressed as linear equations or inequations (i.e., inequality equations or equations with the sign $\geq$ or $\leq$) with the decisive variables. Some simple examples of the formulation of LP problems are as follows.

EXAMPLE 6.1: EXAMPLE OF FORMULATION OF A MAXIMIZATION LP PROBLEM WITH INEQUALITY CONSTRAINTS

A textile unit produces two types of fabrics of equal width—poplin and canvas. Poplin requires 100 g of cotton yarn per meter, whereas canvas requires 160 g of the same cotton yarn per meter. The minimum demand of poplin in the market is 80 meters per day, whereas the demand for canvas is unlimited. The mill has 12,000 g of cotton yarn available per day in its stock. The profit of poplin is Rs. 10 and that for canvas is Rs. 20. The textile mill would like to plan the production to maximize the profit.

To find the objective and constraint functions of this LP problem, let us assume that the mill has to produce x and y meters of poplin and canvas per day, respectively.

Now, if Z is the total profit that the mill would like to maximize, the objective function can be written as

$$\text{Maximize } Z = 10x + 20y \tag{6.1}$$

The constraints of the problem are as follows.

For the amount of yarn available in the mill per day and the demand of poplin fabric, the constraint functions would be

$$120x + 80y \leq 12{,}000 \quad \text{(inequality constraint)} \tag{6.2}$$

$$x \geq 80 \quad \text{(inequality constraint)} \tag{6.3}$$

$$\text{and } x \geq 0; y \geq 0 \quad \text{(non-negativity restrictions)} \tag{6.4}$$

Thus, the problem is to find x and y that will maximize the objective function Z, subject to the inequality constraints and non-negativity restrictions shown in Equations 6.2 through 6.4. Here, x and y are the decisive variables.

EXAMPLE 6.2: EXAMPLE OF FORMULATION OF A MAXIMIZATION LP PROBLEM WITH EQUALITY CONSTRAINTS

Assume that the textile unit mentioned in Example 6.1 produces the same two types of fabrics (poplin and canvas) of equal width. As already mentioned in the earlier example, poplin and canvas require 100 and 160 g of cotton yarn per meter of the fabrics, respectively. The demand of poplin in the market is a minimum of 40 meters per day, whereas the demand for canvas is unlimited. The mill has 12,000 g of cotton yarn available per day in its stock. The profits of poplin and canvas are Rs. 10 and Rs. 20, respectively. The textile mill would like to plan to meet the market demand of poplin, utilize the entire amount of yarn available in the stock, and maximize the profit.

To find the objective and constraint functions of this LP problem, let us assume the mill has to produce x and y meters of poplin and canvas per day, respectively.

Now, if Z is the total profit that the mill would like it to maximize, the objective function can be written as

$$\text{Maximize } Z = 10x + 20y \tag{6.5}$$

The constraints of the above problem are as follows.

For the amount of yarn available in the mill per day and the demand of poplin fabric, the constraint functions can be written as

$$120x + 80y = 12{,}000 \quad \text{(equality constaint)} \tag{6.6}$$

$$x \geq 40 \qquad\qquad\quad \text{(inequality constraint)} \tag{6.7}$$

$$\text{and } x \geq; y \geq 0 \quad\quad \text{(non-negativity restrictions)} \tag{6.8}$$

Thus, the problem is to find x and y that will maximize the objective function Z, subject to equality and inequality constraints shown in Equations 6.6 and 6.7, respectively, and non-negativity restrictions as shown in Equation 6.8. Here, x and y are decisive variables.

EXAMPLE 6.3: EXAMPLE OF FORMULATION OF A MINIMIZATION LP PROBLEM WITH INEQUALITY CONSTRAINTS

A spinning mill uses three types of cotton—A, B, and C—to spin a yarn. The mean fiber length and strength of the mixing must be at least 38.5 mm and 40 gm/tex to achieve the desired yarn strength. The cost, mean length, and strength of the fibers A, B, and C are shown in Table 6.1. The spinning mill desires to minimize the mixing cost of the yarn.

To find the objective and constraint functions of this LP problem, let us assume that the spinning mill mixes cotton fibers A, B, and C in the ratio of $x:y:z$.

TABLE 6.1

Cost, Fiber Properties of Cotton A, B, and C

Cotton Type	Cost/kg (Rs.)	Fiber Mean Length (mm)	Fiber Strength (gm/tex)
A	60	36	37
B	65	38	39
C	70	44	43

Now, the objective of this problem is to minimize the mixing cost. Assuming Z is the mixing cost, the objective function can be written as

$$\text{Minimize } Z = 60x + 65y + 70z \tag{6.9}$$

The constraints of this problem are as follows.
For maintaining mean fiber length of at least 38.5 mm, the constraint function would be

$$36x + 38y + 44z \geq 38.5 \quad \text{(inequality constraint)} \tag{6.10}$$

For maintaining mean fiber strength of at least 40 gm/tex, the constraint function would be

$$37x + 39y + 43y \geq 40 \quad \text{(inequality constraint)} \tag{6.11}$$

The equality constraint and nonnegative restrictions are given in Equations 6.12 and 6.13, respectively:

$$x + y + z = 1 \quad \text{(equality constraint)} \tag{6.12}$$

$$\text{and } x \geq 0, y \geq 0, z \geq 0 \quad \text{(non-negativity restrictions)} \tag{6.13}$$

Thus, the problem is to find x, y, and z that will minimize the objective function Z, subject to the constraints shown in Equations 6.10 through 6.13.

6.2.1 General Form of a Linear Programming Problem

The general LP problem can be written as

$$\text{Maximize or minimize } Z = c_1x_1 + c_2x_2 + c_3x_3 \cdots + c_nx_n \tag{6.14}$$

where Z is the objective function, $x_1, x_2, x_3, \ldots, x_n$ are the decisive variables, and $c_1, c_2, c_3, \ldots, c_n$ are the regression coefficients.
The problem is subject to the following constraints:

$$a_{11}x_1 + a_{12}x_2 + a_{13}x_3 + \cdots + a_{1n}x_n \ (\leq = \geq)b_1$$
$$a_{21}x_1 + a_{22}x_2 + a_{23}x_3 + \cdots + a_{2n}x_n \ (\leq = \geq)b_2$$
$$a_{31}x_1 + a_{32}x_2 + a_{33}x_3 + \cdots + a_{3n}x_n(\leq = \geq)b_3$$
$$\cdots \quad \cdots \quad \cdots \quad \cdots$$
$$a_{m1}x_1 + a_{m2}x_2 + a_{m3}x_3 + \cdots + a_{mn}x_n \ (\leq = \geq)b_m$$

The above set of constraints will involve any one of the following signs: $\leq$, $=$, or $\geq$. The coefficients $a_{11}, a_{12}, \ldots, a_{mn}$ are known as activity parameters, and the constraint quantities $b_1, b_2, ..., b_m$ are known as requirement parameters of the problem.

It is obvious from Equation 6.14 that the objective function is Z, which is either to be minimized or maximized, and the constraints are linear functions of decisive variables.

6.3 Graphical Method

The two most popular methods to solve LP problems are the graphical method and the simplex method. Though the graphical method is comparatively easier, it has restriction in its application (Vohra 2010). The graphical method can be applied only when the objective function involves only two decisive variables. To solve an LP problem with the graphical method, all of the inequations of the constraints are considered as equations, and all of the variables are constrained to be nonnegative. Lines corresponding to these equations are then drawn in a two-dimensional plane. The area thus formed by these lines defines the feasible region or the solution space that satisfies all of the constraints/restrictions simultaneously. Then by trial and error, a point is located in this feasible region with coordinates that will give the optimal value (either maximum or minimum) of the objective function. The optimum point is a corner point of the feasible region (i.e., the vertex of the polygon). As the variables are constrained in non-negativity in all LP programs, only the non-negative quadrant of the two-dimensional plane is examined for optimum values (Panneerselvam 2002).

Examples of application of the graphical method for solving LP problems are as follows.

EXAMPLE 6.4: EXAMPLE OF SOLVING A MAXIMIZATION LP PROBLEM WITH INEQUALITY CONSTRAINTS

Consider the problem explained in Example 6.1 in which a textile unit produces two types of fabrics—poplin and canvas—which have different profit margins and demands in the market. Both of the fabrics require the same cotton yarn of different quantities. The mill has a certain amount of yarn available in its stock. The textile mill would like to plan the production to maximize the profit. The formulation of the above LP problem was discussed in Example 6.1 and is as follows:

$$\text{Maximize} \quad Z = 10x + 20y$$
$$\text{Subject to} \quad 120x + 80y \leq 12,000$$
$$x \geq 80$$
$$\text{and } x \geq 0; y \geq 0$$

where x and y are the lengths of poplin and canvas in meters produced by the mill per day to maximize profit.

To solve this LP problem using the graphical method, initially a graph is constructed by drawing horizontal and vertical axes that represent x and y axes in the *xoy* plane. The x and y on the horizontal and vertical axes represent the lengths of poplin and canvas fabric, respectively. As both x and y are nonnegative, the search for the solution will be restricted to the first quadrant only. Next, the coordinates on the *xoy* plane are computed from the constraints.

Consider the first constraint, $120x + 80y \leq 12{,}000$. Treat it as an equation $120x + 80y = 12{,}000$. From this equation, we get $x = 100$, when $y = 0$; and $y = 150$, when $x = 0$. A line is drawn through the coordinates $(100, 0)$ and $(0, 150)$ as shown in Figure 6.1. It is obvious from the first constraint that any point above this line violates the inequality condition.

Similarly, the second constraint $x \geq 40$ is also plotted on the graph. Here the constraint is a line parallel to the y-axis.

After graphing all of the constraints, the area thus formed, bounded by all of the constraint lines including the boundary points is called the *feasible region* or the *solution space*. The triangle *ABC* shown in Figure 6.1 is the feasible region for the solution.

Since one of the corner points of the feasible region is the optimal value of the objective function, the values of the objective function for each corner point of the triangle are computed:

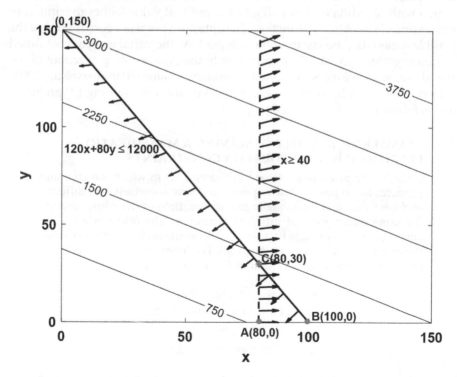

FIGURE 6.1
Feasible region of Example 6.4.

$$Z(A) = 10 \times 80 = 20 \times 0 = 800$$

$$Z(B) = 10 \times 100 + 20 \times 0 = 1,000$$

$$Z(C) = 10 \times 80 + 20 \times 30 = 1,400$$

In this LP problem, the objective function is to be maximized. The solution corresponding to the maximum Z value is selected as the optimum solution. The maximum value of the objective function, that is, $Z = 1,400$, occurs at the coordinate (80, 30).

Hence, the optimal solution of this LP problem is $x = 80$, $y = 30$, and $Z = 1,400$. Thus, the textile mill needs to produce 80 m of poplin and 30 m of canvas per day to maximize the profit, which would be Rs. 1,400.

The MATLAB® coding for the aforesaid maximization problem using the graphical method is given in Section 6.6.

EXAMPLE 6.5: EXAMPLE OF SOLVING MAXIMIZATION LP PROBLEM USING EQUALITY CONSTRAINTS

Consider the problem explained in Example 6.2 in which a textile unit produces poplin and canvas fabrics of equal width. Poplin and canvas require 100 and 160 gm of cotton yarn per meter of fabric, respectively. The mill has 12,000 gm of cotton yarn available in its stock. The profits for poplin and canvas are Rs. 10 and Rs. 20, respectively. The mill wants to maximize its profit subject to meeting the market demand for poplin and canvas and consuming the entire yarn available in the stock. The formulation of this LP problem was already discussed in Example 6.2 and is as follows:

Maximize $\quad Z = 10x + 20y$

Subject to $\quad 120x + 80y = 12,000$

$\qquad\qquad x \geq 40$

$\qquad\qquad$ and $x \geq 0; y \geq 0$

where x and y are the lengths of poplin and canvas produced by the mill per day to maximize profit.

To solve this LP problem, a graph is constructed by drawing horizontal and vertical axes that represent x and y axes in the xoy plane. The x and y on the horizontal and vertical axes represent the lengths of poplin and canvas fabric, respectively. As both x and y are nonnegative, the search for the solution will be restricted to the first quadrant only. Next, the coordinates on the xoy plane are computed from the constraints.

Consider the first constraint, $120x + 80y = 12,000$. From this equation, we get $x = 100$, when $y = 0$; and $y = 150$, when $x = 0$. A line is drawn through the coordinates (100, 0) and (0, 150) as shown in Figure 6.2. As the constraint is equality in nature, the solution lies on the line; any point above or below the line violates the equality condition.

Next, the second constraint $x \geq 40$ is also plotted on the graph. Here the constraint is a line parallel to the y-axis.

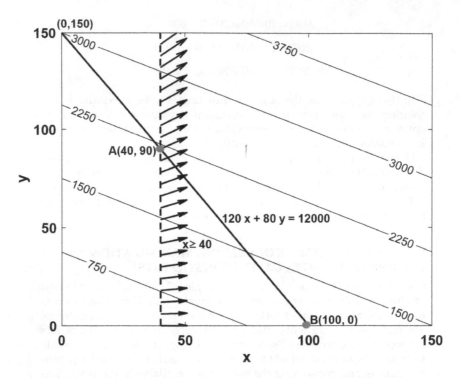

FIGURE 6.2
Feasible region of Example 6.5.

After plotting all of the constraints, the feasible region of the objective function can be located. It is obvious from the graph that the equality constraint line, bounded by the inequality constraint lines including the boundary points is the feasible region or solution space.

The line segment AB shown in Figure 6.2, is the feasible region for the solution. Since the optimal value of the objective function occurs at one of the extreme points of the feasible region, the objective function value for each of the extreme points of the line segment AB is computed:

$$Z(A) = 10 \times 40 + 20 \times 90 = 2,200$$

$$Z(B) = 10 \times 100 + 20 \times 0 = 1,000$$

In this LP program, the objective function, profit, should be maximized. The solution corresponding to the maximum profit is selected as the optimum solution. The maximum value of the objective function, $Z = 2,200$, occurs at the coordinate (40, 90). Hence, the optimum solution is $x = 40$, $y = 90$, and $Z = 2,200$. Therefore, the textile mill should produce 40 m of poplin and 90 m of canvas per day to maximize the profit, which would be Rs. 2,200, therefore satisfying all of the constraints.

The MATLAB® coding for the aforesaid optimization problem using graphical method is given in Section 6.6.

EXAMPLE 6.6: EXAMPLE OF SOLVING A MINIMIZATION LP PROBLEM USING INEQUALITY CONSTRAINTS

A spinning mill manufactures yarn by mixing two types of cotton fiber—*A* and *B*. To achieve the desired yarn quality, the mean fiber length and strength of the mixing should be at least 40 mm and 44 gm/tex, respectively. The cost, mean length, and strength of the fibers *A* and *B* are given in Table 6.2. The spinning mill wants to minimize the mixing cost of the yarn.

Now, the objective of this problem is to minimize the mixing cost. Let us assume the spinning mill mixes cotton fibers *A* and *B* in the ratio *x:y*. If *Z* is the mixing cost, this problem can be written as

$$\text{Minimize} \quad Z = 60x + 80y$$
$$\text{Subject to} \quad 36x + 42y \geq 40$$
$$38x + 46y \geq 44 \tag{6.15}$$
$$x + y = 1$$
$$x \geq 0, y \geq 0$$

The problem is to find *x* and *y* that will minimize the objective function *Z*, subject to the constraints shown in Equation 6.15.

Similar to previously discussed problems, initially a graph is drawn along the horizontal and vertical axes that represents *x* and *y* axes in the *xoy* plane. The *x* and *y* on the horizontal and vertical axes represent the quantities of fibers *A* and *B*, respectively. As both the *x* and *y* are nonnegative, the search for the solution will be restricted to the first quadrant only. Then, the coordinates on the *xoy* plane are computed from the constraints.

The first constraint is treated as an equation, $36x + 42y = 40$, from which we get $x = 1.11$, when $y = 0$; and $y = 0.95$, when $x = 0$. The first constraint line is drawn through the coordinates (1.11, 0) and (0, 0.95).

Similarly, constraints $38x + 46y \geq 44$ and $x + y = 1$ are also plotted on the graph. The extreme points for the objective functions are *A* and *B* as shown in Figure 6.3. Consequently, the objective function value for *A* and *B* is computed:

$$Z(A) = 60 \times 0 + 80 \times 1 = 80$$

$$Z(B) = 60 \times 0.25 + 80 \times 0.75 = 75$$

TABLE 6.2

Cost and Fiber Properties of Cotton *A* and *B*

Cotton Type	Cost/kg (Rs.)	Fiber Mean Length (mm)	Fiber Strength (gm/tex)
A	60	36	38
B	80	42	46

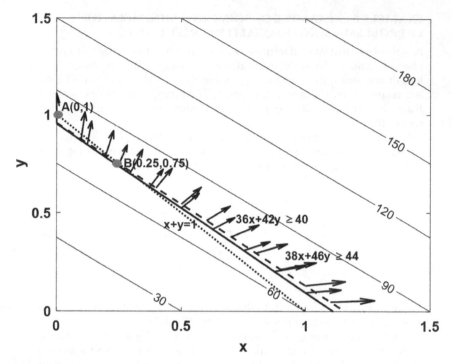

FIGURE 6.3
Feasible region of Example 6.6.

Since, in this LP problem, the objective function is to be minimized, the solution corresponding to the minimum Z value is selected as the optimum solution. The minimum value of the objective function, $Z = 75$, occurs at the coordinate (0.25, 0.75).

Hence, the optimal solution is $x = 0.25$, $y = 0.75$, and $Z = 75$. To minimize the mixing cost, the spinning mill should mix fibers A and B in the ratio of 1:3 (0.25:075 = 1:3); thus, the minimum mixing cost per kilogram would be Rs. 75.

The MATLAB® coding for the aforesaid optimization problem using the graphical method is given in Section 6.6.

6.4 Simplex Method

In the graphical method of solving LP problems, initially the feasible region is identified using the constraint lines and then an optimal solution is obtained from the extreme points of this feasible region. Though this method is efficient and easy to understand, it suffers from a great limitation in that it can be used for linear problems involving only two decisive variables. In real-world situations, often more than two decisive variables are involved, which cannot be handled by the graphical method. So, for such linear problems, the *simplex method* can be used, which is found to be efficient in solving LP problems of any magnitude (Deb 2010; Raju 2009).

The simplex algorithm, proposed by G. B. Dantzig (1963), is an iterative approach (step by step) for finding the optimal solution of a LP problem. The concept of the simplex method is like that of the graphical method in which extreme points of a multidimensional figure (n-dimensional polyhedron) are searched for an optimal solution. The method starts with a basic level, and at each iteration (extreme point) it projects the improvement in the objective function over its earlier step. A solution is considered as optimal when no further improvement of the objective function is possible (Panneerselvam 2002; Raju 2009).

The simplex method is demonstrated using a suitable numeric problem.

EXAMPLE 6.7: MAXIMIZATION LP PROBLEM

Consider the following maximization LP problem

$$\begin{aligned} \text{Maximize} \quad & Z = 10x + 20y \\ \text{Subject to} \quad & x + y \leq 40 \\ & 4x + 8y \leq 80 \\ & \text{and } x \geq 0;\, y \geq 0 \end{aligned} \tag{6.16}$$

Step 1: Standardization of LP problem

   a. The LP problem should be a maximization type. In case the LP is of a minimization type, both sides of the objective function are multiplied by −1, which converts the LP to a maximization type.

   For an example, if the objective function is Minimize $Z = 60x + 80y$, then it should be written as Maximize $Z' = -60x - 80y$, where $Z' = -Z$.

   The problem in Example 6.7 is a maximization-type problem.

   b. The constraint values should be ≥ 0. In case the constraint values are ≤ 0, both sides of the constraint should be multiplied by −1, such that all constraint values become ≥ 0.

   For example, if the constraint is $10x + 8y \leq -120$, then it should be written as $-10x - 8y \geq 120$.

   The values of both constraints in Example 6.7 are ≥ 0.

   c. The inequality constraints are converted into equality form by adding *slack, surplus,* or *artificial* variables. Slack variables are added if the constraint is of the "≤" type. Surplus variables are subtracted and artificial variables are added for constraints of the "≥" type. Only artificial variables should be added for those constraints with "=" signs.

The standard form of the LP problem of Example 6.7 is as follows:

$$\text{Maximize} \quad Z = 10x + 20y + 0.S_1 + 0.S_2$$
$$\text{Subject to} \quad x + y + 1.S_1 + 0.S_2 = 40$$
$$4x + 8y + 0.S_1 + 1.S_2 = 80$$
$$x, y, S_1 \text{ and } S_2 \geq 0$$

where S_1 and S_2 are slack variables.

The standard form of LP is known as the *canonical* form. A canonical form is the form in which each constraint of a LP has a basic variable. A variable is said to be a basic variable if its coefficient is 1 in one of the constraints and 0 in the remaining constraints. If all of the constraints are of the "≤" type, then the standard form would be in canonical form.

In this example, the basic variables are S_1 and S_2. The coefficients of S_1 and S_2 in the first constraint are 1 and 0, respectively, whereas in the second constraint the coefficients are 0 and 1, respectively.

Step 2: Finding the Initial Basic Feasible Solution

In an LP problem with n variables (including all decision, slack, surplus, and artificial variables) and m simultaneous linear equations ($n > m$), the solution obtained for m variables by setting ($n - m$) variables equal to zero is known as the basic solution. The ($n - m$) variables that are set equal to zero are known as nonbasic variables. The other m variables whose values are obtained by solving the equations are the basic variables.

The initial basic feasible solution (IBFS) to an LP problem is the solution for which the values of m are all greater than or equal to zero. In Example 6.7, the IBFS is $S_1 = 40$ and $S_2 = 80$ (when $x = y = 0$ and $S_1, S_2 \geq 0$).

Step 3: Setup of Initial Simplex Table

The initial simplex table for the given problem is constructed as shown in Table 6.3, where C_j is the coefficient of the jth variable of the objective function; CB_i is the coefficient of the ith basic variable in the objective function; Z_j is the summation of products of CB_i and a_{ij} (coefficient for the ith row and jth column of the table); and ($C_j - Z_j$) is the net contribution.

Now for a maximization problem, optimality is reached if all of the values of ($C_j - Z_j$) are equal to or less than zero. In Table 6.3, all of the values for ($C_j - Z_j$) are neither equal to nor less than zero; hence, optimality is not reached, and the solution can be further improved.

Step 4: Finding "Exiting" and "Entering" Variables

As all of the values for ($C_j - Z_j$) are neither equal to nor less than zero, optimality is not reached, and the solution can be improved further in the next iteration. In the initial simplex table, the column with maximum ($C_j - Z_j$) is chosen. This column is known as the *key column*, and the variable in the key column is called the *entering variable*. To maintain

TABLE 6.3

Initial Simplex Table (Example 6.7)

CB_i	C_j Basic Variable (BV)	10 x	20 y	0 S_1	0 S_2	Solution	Ratio
0	S_1	1	1	1	0	40	$\dfrac{40}{1}=40$
0	S_2	4	8	0	1	80	$\dfrac{80}{8}=10^a$
	Z_j	0	0	0	0	0	
	$C_j - Z_j$	10	20^b	0	0		

a Key row; 8 is the key element.
b Key column.

the feasibility of the solution in each iteration, the ratio of *solution* and *value in the key column* is calculated. The basic variable corresponding to the minimum ratio is called the *exiting variable*, and the corresponding row is called the *key row*. The value at the intersection of the key column and the key row is known as the *key element*. In Table 6.3, the maximum $(C_j - Z_j)$ is 20. The shaded column is the key column, and the entering variable is y. The ratio of the solution and the value in the key column is calculated. The minimum ratio is 10. The shaded column is the key row, and the exiting variable is S_2. The key element is 8.

Step 5: Rewriting the Next Iteration Table

The exiting basic variable is replaced by the entering variable along with its coefficients. The values of the rows in the new table are computed. The key element is transformed as unity. Therefore, the total key row is divided by the key element. The values for other rows are computed using the following formula:

$$\text{New value} = \text{Old value} - \frac{\text{Key column value} \times \text{Key row value}}{\text{Key value}}$$

For example, the computation of the new value of row 1 and column 1 is as follows:

$$\text{New value} = 1 - \frac{1 \times 4}{8} = \frac{1}{2}$$

The next iteration table with the new value is shown in Table 6.4.

In Table 6.4, all of the values for $(C_j - Z_j)$ are either equal to or less than zero. Hence, the optimality is reached and the corresponding optimal solution is as follows:

$$x = 0;\ y = 10;\ Z\ (\text{optimum}) = 200$$

TABLE 6.4

Iteration Table 1 (Example 6.7)

C_j		10	20	0	0	
CB_i	Basic Variable (BV)	x	y	S_1	S_2	Solution
0	S_1	$\dfrac{1}{2}$	0	1	$-\dfrac{1}{8}$	30
20	y	$\dfrac{1}{2}$	1	0	$\dfrac{1}{8}$	10
	Z_j	10	20	0	$\dfrac{5}{2}$	200
	$C_j - Z_j$	0	0	0	$-\dfrac{5}{2}$	

6.4.1 Big M Method

In the earlier example, the constraints were of the "$\leq$" type, and slack variables were added to the constraints to convert their inequality form into an equality form. When an LP problem has constraints of "$=$" or "$\geq$", a new variable called an *artificial variable* is incorporated with a positive coefficient (Panneerselvam 2002). For an objective function of maximization or minimization type, the coefficient of the artificial variable in the objective function should be "$-M$" or "$+M$," respectively, where M is a very large value associated with the artificial variables. The Big M method is demonstrated with the following suitable numeric examples.

EXAMPLE 6.8: MAXIMIZATION LINEAR PROBLEM

Consider the following maximization linear problem in which constraints are of "$=$" and "$\geq$" types:

$$\begin{aligned} \text{Maximize} \quad & Z = 10x + 20y \\ \text{Subject to} \quad & 120x + 80y = 12{,}000 \\ & x \geq 40 \\ & x \geq, y \geq 0 \end{aligned} \qquad (6.17)$$

The canonical form of this model is presented as follows after incorporating slack and artificial variables:

$$\begin{aligned} \text{Maximize} \quad & Z = 10x + 20y + 0.S_1 - MR_1 - MR_2 \\ \text{Subject to} \quad & 120x + 80y + R_1 = 12{,}000 \\ & x - S_1 + R_2 = 40 \\ & x, y, S_1, R_1 \text{ and } R_2 \geq 0; M \text{ is large value} \end{aligned}$$

where S_1 is a slack variable, and R_1 and R_2 are artificial variables.

TABLE 6.5

Initial Simplex Table (Example 6.8)

	C_j	10	20	0	$-M$	$-M$		
CB_i	Basic Variable (BV)	x	y	S_1	R_1	R_2	Solution	Ratio
$-M$	R_1	120	80	0	1	0	12,000	$\dfrac{12,000}{120} = 100$
$-M$	R_2	1	0	-1	0	1	40	$\dfrac{40}{1} = 40^a$
	Z_j	$-121M$	$-80M$	M	$-M$	$-M$	$-12,040M$	
	$C_j - Z_j$	$10 + 121M^b$	$20 + 80M$	$-M$	0	0		

[a] Key row; 1 is the key element.
[b] Key column.

The initial simplex table for the given problem is constructed as discussed earlier in Example 6.7 and is shown in Table 6.5.

In Table 6.5, all of the values for $(C_j - Z_j)$ are neither zero nor less than zero. Therefore, optimality is not reached, and the solution can be improved further. In the table, the maximum $(C_j - Z_j)$ is $10 + 121M$. The shaded column is the key column, and the entering variable is x. To maintain the feasibility of the solution in each iteration, the ratio of *solution* and *value in the key column* is calculated. The minimum ratio is 40. The shaded row is the key row, and the exiting variable is R_2. The key element is 1.

The exiting variable is replaced by the entering variable along with its coefficients. The values of the rows in the new table are computed as discussed in Example 6.7. It is to be noted that when an artificial variable (R_2 in this case) is replaced, its column should be deleted from the table for further iteration. So, the column of R_2 is deleted.

The next iteration table with a new value is shown in Table 6.6.

TABLE 6.6

Iteration Table 1 (Example 6.8)

	C_j	10	20	0	$-M$		
CB_i	Basic Variable (BV)	x	y	S_1	R_1	Solution	Ratio
$-M$	R_1	0	80	120	1	7,200	$\dfrac{7,200}{120} = 60^a$
10	x	1	0	-1	0	40	$\dfrac{40}{-1} = -40$ (ignore)
	Z_j	10	$-80M$	$-120M - 10$	$-M$	$400 - 7,200M$	
	$C_j - Z_j$	0	$80M + 20$	$120M + 10^b$	0		

[a] Key row; 120 is the key element.
[b] Key column.

TABLE 6.7

Iteration Table 2 (Example 6.8)

CB$_i$		C_j	10	20	0		
	Basic Variable (BV)		x	y	S_1	Solution	Ratio
0	S_1		0	$\frac{2}{3}$	1	60	$\frac{60}{2/3}=90^a$
10	x		1	$\frac{2}{3}$	0	100	$\frac{100}{2/3}=150$
	Z_j		10	$\frac{20}{3}$	0	1,000	
	$C_j - Z_j$		0	$\frac{40}{3}^b$	0		

[a] Key row; $\frac{2}{3}$ is the key element.
[b] Key column.

In Table 6.6, all of the values of $(C_j - Z_j)$ are neither equal to nor less than zero. Hence, optimality is not reached. The maximum $(C_j - Z_j)$ is 120M + 10. So, the entering variable is S_1. The minimum ratio is 60 (ratio with indefinite number and with negative sign should be ignored). So, the exiting variable is R_1. The column of R_1 is deleted. The values of the rows are computed. The next iteration table with new values is shown in Table 6.7.

In Table 6.7, the maximum value of $(C_j - Z_j)$ is 40/3. So, the entering variable is y. The minimum ratio is 90. So, the exiting variable is S_1. The value of the rows of the next iteration table is computed and is shown in Table 6.8.

In Table 6.8, all of the values for $(C_j - Z_j)$ are zero and less than zero. Hence, the optimality is reached, and the corresponding optimal solution is as follows:

$$x = 40; y = 90 \quad \text{and} \quad Z \text{ (max)} = 2,200$$

TABLE 6.8

Iteration Table 4 (Example 6.8)

CB$_i$		C_j	10	20	0	
	Basic Variable (BV)		x	y	S_1	Solution
20	y		0	1	$\frac{3}{2}$	90
10	x		1	0	-1	40
	Z_j		10	20	20	2,200
	$C_j - Z_j$		0	0	-20	

EXAMPLE 6.9: SIMPLEX METHOD TO SOLVE LP MINIMIZATION PROBLEM

Consider the mixing problem in Example 6.6 in which the spinning mill desires to minimize the mixing cost of yarn. The LP problem is given as follows and is solved by the simplex method:

$$
\begin{aligned}
\text{Minimize} \quad & Z = 60x + 80y \\
\text{Subject to} \quad & 36x + 42y \geq 40 \\
& 38x + 46y \geq 44 \\
& x + y = 1 \\
& x \geq 0, y \geq 0
\end{aligned}
$$

Since the LP is of the minimization type, both sides of the objective function are multiplied by -1, which converts the LP to the maximization type. The program can be written as

$$
\begin{aligned}
\text{Maximize} \quad & Z' = -Z = -60x - 80y \\
\text{Subject to} \quad & 36x + 42y \geq 40 \\
& 38x + 46y \geq 44 \\
& x + y = 1 \\
& x \geq 0, y \geq 0
\end{aligned}
$$

The canonical form of this model is presented as follows after incorporating slack (S_1 and S_2) and artificial variables (R_1, R_2 and R_3):

$$
\begin{aligned}
\text{Maximize} \quad & Z' = -60x - 80y + 0.S_1 + 0.S_2 - MR_1 - MR_2 - MR_3 \\
\text{Subject to} \quad & 36x + 42y - S_1 + R_1 = 40 \\
& 38x + 46y - S_2 + R_2 = 44 \\
& x + y + R_3 = 1 \\
& x, y, S_1, S_2, R_1, R_2 \text{ and } R_3 \geq 0; \text{ M is a large value}
\end{aligned}
$$

TABLE 6.9

Initial Simplex Table (Example 6.9)

	C_j	-60	-80	0	0	$-M$	$-M$	$-M$		
CB_i	Basic Variable (BV)	x	y	S_1	S_2	R_1	R_2	R_3	Solution	Ratio
$-M$	R_1	36	42	-1	0	1	0	0	40	$\frac{40}{42} = 0.9524$[a]
$-M$	R_2	38	46	0	-1	0	1	0	44	$\frac{44}{46} = 0.9565$
$-M$	R_3	1	1	0	0	0	0	1	1	$\frac{1}{1} = 1$
	Z'_j	$-75M$	$-89M$	M	M	$-M$	$-M$	$-M$	0	
	$C_j - Z'_j$	$75M - 60$	$89M - 80$[b]	$-M$	$-M$	0	0	0		

[a] Key row; 42 is the key element.
[b] Key column.

TABLE 6.10

Iteration Table 1 (Example 6.9)

CB$_i$	C_j Basic Variable (BV)	-60 x	-80 y	0 S_1	0 S_2	$-M$ R_2	$-M$ R_3	Solution	Ratio
-80	y	$\dfrac{6}{7}$	1	$-\dfrac{1}{42}$	0	0	0	$\dfrac{20}{21}$	$\dfrac{20/21}{-1/42}=-40$ (ignore)
$-M$	R_2	$-\dfrac{10}{7}$	0	$\dfrac{23}{21}$	-1	1	0	$\dfrac{4}{21}$	$\dfrac{4/21}{23/21}=0.1739^{a}$
$-M$	R_3	$\dfrac{1}{7}$	0	$\dfrac{1}{42}$	0	0	1	$\dfrac{1}{21}$	$\dfrac{1/21}{1/42}=2$
	Z'_j	$\dfrac{9M}{7}-\dfrac{480}{7}$	-80	$-\dfrac{47M}{42}+\dfrac{40}{21}$	M	$-M$	$-M$	$-\dfrac{1,600}{21}$	
	$C_j-Z'_j$	$\dfrac{60}{7}-\dfrac{9M}{7}$	0	$\dfrac{47M}{42}-\dfrac{40}{21}^{\,b}$	$-M$	0	0		

[a] Key row; $\dfrac{23}{21}$ is the key element.
[b] Key column.

The initial simplex table for the given problem is constructed and shown in Table 6.9.

It is apparent from Table 6.9 that all the $(C_j - Z'_j)$ are neither equal to nor less than zero. Therefore, optimality is not reached. The key column, ratio of *solution* and *value in the key column*, key row, and key element are evaluated and shown in the same table. The key column and key rows

TABLE 6.11

Iteration Table 2 (Example 6.9)

CB$_i$	C_j Basic Variable (BV)	-60 x	-80 y	0 S_1	0 S_2	$-M$ R_3	Solution	Ratio
-80	y	$\dfrac{19}{23}$	1	0	$-\dfrac{1}{46}$	0	$\dfrac{22}{23}$	$\dfrac{22/23}{19/23}=1.1579$
0	S_1	$-\dfrac{30}{23}$	0	1	$-\dfrac{21}{23}$	0	$\dfrac{4}{23}$	$\dfrac{4/23}{-30/23}=-0.12$ (ignore)
$-M$	R_3	$\dfrac{4}{23}$	0	0	$\dfrac{1}{46}$	1	$\dfrac{1}{23}$	$\dfrac{1/23}{4/23}=0.25^{a}$
	Z'_j	$-\dfrac{4M}{23}-\dfrac{1,520}{23}$	-80	0	$-\dfrac{M}{46}+\dfrac{40}{23}$	$-M$	$-\dfrac{1,760}{23}$	
	$C_j-Z'_j$	$\dfrac{4M}{23}+\dfrac{140}{23}^{\,b}$	0	0	$\dfrac{M}{46}-\dfrac{40}{23}$	0		

[a] Key row; $\dfrac{4}{23}$ is the key element.
[b] Key column.

TABLE 6.12

Iteration Table 3 (Example 6.9)

CB_i	C_j Basic Variable (BV)	-60 x	-80 y	0 S_1	0 S_2	Solution
-80	y	0	1	0	$-\dfrac{1}{8}$	$\dfrac{3}{4}$
0	S_1	0	0	1	$-\dfrac{3}{4}$	$\dfrac{1}{2}$
-60	x	1	0	0	$\dfrac{1}{8}$	$\dfrac{1}{4}$
	Z_j'	-60	-80	0	$\dfrac{5}{2}$	-75
	$C_j - Z_j'$	0	0	0	$-\dfrac{5}{2}$	

are shaded in gray. Considering that the entering variable is y and exiting variable is R_1, the corresponding key operations are shown in Table 6.10.

In Table 6.10 the optimality is not reached. With the entering variable as S_1 and exiting variable as R_2, the corresponding key operations are as shown in Table 6.11.

It is apparent from Table 6.11 that optimality is not even reached. So, considering x as the entering variable and R_3 as the exiting variable, corresponding key operations are shown in Table 6.12.

In Table 6.12, all of the values for $(C_j - Z_j')$ are either less than or equal to zero. Hence, the optimality is reached, and the corresponding optimal solution is as follows:

$$x = 1/4; y = 3/4 \quad \text{and} \quad Z' (\text{max}) = -75$$

Therefore, Z (min) $= -Z'(\text{max}) = 75$.

6.4.2 Two-Phase Method

In the Big M method, the value of M is assumed to be a very large number that may lead to misleading results due to the rounding of coefficients while carrying out different iterations. The two-phase method overcomes these limitations for problems with "$\geq$" or "$=$" constraints (Vohra 2010; Panneerselvam 2002). This method has two different phases to solve a LP problem and, hence, such nomenclature. In the initial phase, the sum of the artificial variable is minimized to obtain a basic feasible solution. In the next phase, the original objective function is minimized starting with the basic feasible solution obtained in the initial phase. The two-phase method is discussed herewith with suitable numeric examples.

EXAMPLE 6.10: TWO-PHASE SIMPLEX METHOD TO SOLVE LP PROBLEM

Reconsider the mixing problem in Example 6.6 in which the spinning mill wants to minimize the mixing cost of yarn. The LP problem is as follows and is solved by the two-phase simplex method:

$$\text{Minimize} \quad Z = 60x + 80y$$
$$\text{Subject to} \quad 36x + 42y \geq 40$$
$$38x + 46y \geq 4,400$$
$$x + y = 1$$
$$x \geq 0, y \geq 0$$

Phase I

Step 1

In case an LP problem has only "≤" type constraints, Phase II can be directly used to solve the problem. In Example 6.10, the constraints are of "≥" and "=" type.

Step 2

The LP problem is converted into a maximization form. Surplus variables (S_1 and S_2) and artificial variables (R_1, R_2, and R_3) are added to the constraints. The problem becomes

$$\text{Maximize} \quad Z' = -Z = -60x - 80y$$
$$\text{Subject to} \quad 36x + 42y - S_1 + R_1 = 40$$
$$38x + 46y - S_2 + R_2 = 44$$
$$x + y + R_3 = 1$$
$$x, y, S_1, S_2, R_1, R_2 \text{ and } R_3 \geq 0$$

Step 3

Assigning zero coefficients to each of the decisive variables (x and y in this example) and surplus variables (S_1 and S_2 in this example), and a −1 coefficient to each of the artificial variables (R_1, R_2, and R_3 in this example) in the objective function, the LP problem becomes

$$\text{Maximize} \quad Z' = -Z = -R_1 - R_2 - R_3$$
$$\text{Subject to} \quad 36x + 42y - S_1 + R_1 = 40$$
$$38x + 46y - S_2 + R_2 = 44$$
$$x + y + R_3 = 1$$
$$x, y, S_1, S_2, R_1, R_2 \text{ and } R_3 \geq 0$$

Step 4

Prepare the initial simplex table for Phase I. The initial simplex table for Phase I of the given problem is constructed and shown in Table 6.13.

TABLE 6.13

Phase I—Initial Simplex Table (Example 6.10)

CB$_i$	C_j Basic Variable (BV)	0 x	0 y	0 S_1	0 S_2	-1 R_1	-1 R_2	-1 R_3	Solution	Ratio
-1	R_1	36	42	-1	0	1	0	0	40	$\frac{40}{42}=0.9524^a$
-1	R_2	38	46	0	-1	0	1	0	44	$\frac{44}{46}=0.9565$
-1	R_3	1	1	0	0	0	0	1	1	$\frac{1}{1}=1$
	Z_j'	-75	-89	1	1	-1	-1	-1	0	
	C_j-Z_j'	75	89^b	-1	-1	0	0	0		

^{a} Key row; 42 is the key element.
^{b} Key column.

Step 5

The usual simplex method is applied until the optimality is reached. It is apparent from Table 6.13 that all the $(C_j - Z_j')$ are neither equal to nor less than zero. Therefore, optimality is not reached. The key column, ratio of *solution* and *value in the key column*, key row, and key element are evaluated and shown in the same table. The key column and key rows are shaded in gray. Considering that the entering variable is y and the exiting variable is R_1, the corresponding key operations are shown in Table 6.14.

TABLE 6.14

Phase I—Iteration Table 1 (Example 6.10)

CB$_i$	C_j Basic Variable (BV)	0 x	0 y	0 S_1	0 S_2	-1 R_1	-1 R_2	-1 R_3	Solution	Ratio
0	y	$\frac{6}{7}$	1	$-\frac{1}{42}$	0	$\frac{1}{42}$	0	0	$\frac{20}{21}$	$\frac{20/21}{-1/42}=-40$ (ignore)
-1	R_2	$-\frac{10}{7}$	0	$\frac{23}{21}$	-1	$-\frac{23}{21}$	1	0	$\frac{4}{21}$	$\frac{4/21}{23/21}=0.1739^a$
-1	R_3	$\frac{1}{7}$	0	$\frac{1}{42}$	0	$-\frac{1}{42}$	0	1	$\frac{1}{21}$	$\frac{1/21}{1/42}=2$
	Z_j'	$\frac{9}{7}$	0	$-\frac{47}{42}$	1	$\frac{47}{42}$	-1	-1	0	
	C_j-Z_j'	$-\frac{9}{7}$	0	$\frac{47}{42}^b$	-1	$-\frac{89}{42}$	0	0		

^{a} Key row; $\frac{23}{21}$ is the key element.
^{b} Key column.

TABLE 6.15

Phase I—Iteration Table 2 (Example 6.10)

C_j		0	0	0	0	−1	−1	−1		
CB_i	Basic Variable (BV)	x	y	S_1	S_2	R_1	R_2	R_3	Solution	Ratio
0	y	$\dfrac{19}{23}$	1	0	$-\dfrac{1}{46}$	0	$\dfrac{1}{46}$	0	$\dfrac{22}{23}$	$\dfrac{22/23}{19/23}=1.1579$
0	S_1	$-\dfrac{30}{23}$	0	1	$-\dfrac{21}{23}$	−1	$\dfrac{21}{23}$	0	$\dfrac{4}{23}$	$\dfrac{4/23}{-30/23}=-0.13$ (ignore)
−1	R_3	$\dfrac{4}{23}$	0	0	$\dfrac{1}{46}$	0	$-\dfrac{1}{46}$	1	$\dfrac{1}{23}$	$\dfrac{1/23}{4/23}=0.25^a$
	Z_j'	$-\dfrac{4}{23}$	0	0	$-\dfrac{1}{46}$	0	$\dfrac{1}{46}$	−1	0	
	$C_j - Z_j'$	$\dfrac{4}{23}^{\,b}$	0	0	$\dfrac{1}{46}$	1	$-\dfrac{47}{46}$	0		

^a ̶ see below

a Key row; $\dfrac{4}{23}$ is the key element.
b Key column.

In Table 6.14 the optimality is not reached. With the entering variable as S_1 and exiting variable as R_2, the corresponding key operations are shown in Table 6.15. It is apparent from Table 6.15 that optimality is not even reached. So, considering x as the entering variable and R_3 as the exiting variable, the corresponding key operations are shown in Table 6.16.

In Table 6.16, all of the values of $(C_j - Z_j') \le 0$; hence, optimality is reached, and the corresponding optimal solution is as follows:

$$x = 1/4; y = 3/4 \quad \text{and} \quad Z'(\max) = 0$$

TABLE 6.16

Phase I—Iteration Table 3 (Example 6.10)

C_j		0	0	0	0	−1	−1	−1	
CB_i	Basic Variable (BV)	x	y	S_1	S_2	R_1	R_2	R_3	Solution
0	Y	0	1	0	$-\dfrac{1}{8}$	0	$\dfrac{1}{8}$	$-\dfrac{19}{4}$	$\dfrac{3}{4}$
0	S_1	0	0	1	$-\dfrac{3}{4}$	−1	$\dfrac{3}{4}$	$\dfrac{15}{2}$	$\dfrac{1}{2}$
0	x	1	0	0	$\dfrac{1}{8}$	0	$-\dfrac{1}{8}$	$\dfrac{23}{4}$	$\dfrac{1}{4}$
	Z_j'	0	0	0	0	0	0	0	0
	$C_j - Z_j'$	0	0	0	0	−1	−1	−1	

In Table 6.16, the objective function is zero, and the set of basic variables does not contain any artificial variables; therefore, the given problem has a feasible solution and Phase II can be performed.

Phase II

Step 6

A modified table is formed for Phase II. The optimal simplex table of Phase I can be used as the initial simplex table for Phase II. The columns of the artificial variables that are currently nonbasic should be eliminated. Necessary changes are made in the CB_i column by substituting the actual coefficients to the variables in the objective function. In case some artificial variables are present at zero level in the basic solution of the optimal table in Phase I, its objective functions coefficients should be substituted with zero. Further iterations are carried out until the optimality is reached.

The iteration of Phase II is shown in Table 6.17. The columns of the artificial variables (R_1, R_2, and R_3) are eliminated. Necessary changes are made in the CB_i column by substituting the actual coefficients to the variables in the objective function (i.e., CB_i of y and x are substituted with -80 and -60 as given in the objective function, i.e., Maximize $Z' = -60x - 80y$). It is obvious from Table 6.17 that all of the values for $(C_j - Z'_j)$ are either equal to or less than zero; hence, the optimality is reached. Table 6.17 gives an optimal solution, and no further iterations are required. The corresponding optimal solution is as follows:

$$x = 1/4; y = 3/4 \quad \text{and} \quad Z' \text{ (max)} = -75$$

Therefore, Z (min) $= -Z'(\text{max}) = 75$.

TABLE 6.17

Phase II—Iteration Table 1 (Example 6.10)

CB_i	C_j Basic Variable (BV)	0 x	0 y	0 S_1	0 S_2	Solution
-80	y	0	1	0	$-\dfrac{1}{8}$	$\dfrac{3}{4}$
0	S_1	0	0	1	$-\dfrac{3}{4}$	$\dfrac{1}{2}$
-60	x	1	0	0	$\dfrac{1}{8}$	$\dfrac{1}{4}$
	Z'_j	-60	-80	0	$\dfrac{5}{2}$	-75
	$C_j - Z'_j$	0	0	0	$-\dfrac{5}{2}$	

6.5 Applications

There are few reported works on the application of LP in the domain of textile manufacturing. LaFerney (1969) demonstrated the application of the LP technique determining a blend ratio of fiber types to maximize profit with prerequisite processing performance and product quality in a typical textile mill. Lee and Kin (1984) and Kang and Lee (1988) have shown how the LP approach can be used to decide the blend ratio of cotton and wool fibers for optimum yarn properties and spinning performance. Gupta et al. (2006) applied an algorithm based on the LP approach to maximize the number of individuals that can be accommodated into a predetermined number of garment sizes from a given population. Majumdar et al. (2011) cited an example of optimization of a product mix problem in a functional clothing manufacturing unit using both graphical and simplex methods of LP. Tesfaye et al. (2016) devised an efficient resource utilization mechanism for the Ethiopian apparel sector to improve their resource utilization and profitability using an LP technique. Harianto (2018) optimized woven fabric production to maximize the business profit of a weaving plant (PT. Argo Pantes, Tangerang) using the LP approach.

6.6 MATLAB® Coding

MATLAB® Coding for the Optimization Problem Given in Example 6.4 Using Graphical Method

```
clc
clear all
close all
format short g
 % Maximize 10x+ 20y
% Subject to
% 120x+ 80y <= 12000
% x>=80
% x, y>=0
 f=[-10 -20];
b=[12000 -80];
A=[120 80;-1 0];
lb=[0 0];
ub=[];
Aeq=[];
beq=[];
[x fn_value]=linprog(f,A,b,Aeq,beq,lb,ub)
x1=x(1);
x2=x(2);
```

```
Obj_fn= 10*x1 + 20*x2
Cons_1=120*x1 + 80*x2
Cons_2=x1 + 0*x2
```

MATLAB® Coding for the Optimization Problem Given in Example 6.5 Using Graphical Method

```
clc
clear all
close all
format short g
 % Maximize 10x+ 20y
% Subject to
% 120x+ 80y = 12000
% x>=40
% x, y>=0
 f=[-10 -20];
b=[-40];
A=[-1 0];
lb=[0 0];
ub=[];
Aeq=[120 80];
beq=[12000];
[x fn_value]=linprog(f,A,b,Aeq,beq,lb,ub)
x1=x(1);
x2=x(2);
Obj_fn= 10*x1 + 20*x2
Cons_1=120*x1 + 80*x2
Cons_2=x1 + 0*x2
```

MATLAB® Coding for the Optimization Problem Given in Example 6.6 Using Graphical Method

```
clc
clear all
close all
format short g
 % Maximize 60x+ 80y
% Subject to
% 36x+ 42y >= 40
% 38x+ 46y >=44
% x + y = 1
% x, y >=0
 f=[60 80];
b=[-40 -44];
A=[-36 -42;-38 -46];
lb=[0 0];
ub=[];
```

```
Aeq=[1 1];
beq=[1];
[x fn_value]=linprog(f,A,b,Aeq,beq,lb,ub)
x1=x(1);
x2=x(2);
Obj_fn= 60*x1 + 80*x2
Cons_1=36*x1 + 42*x2
Cons_2=38*x1 + 46*x2
cons_3=x1+x2
```

6.7 Summary

LP methods are some of the most popular, versatile, and commonly used approaches that provide optimal solutions to the problems with linear objective and constraint functions. This chapter presents a detailed application of graphical and simplex methods of solving linear programs with suitable examples related to textile manufacturing, along with coding in MATLAB® language. In the graphical method, initially a feasible region is identified in a two-dimensional plane, after satisfying all of the constraints simultaneously. Then, by a trial-and-error method, a point is located in the feasible region whose coordinate gives the optimum value. Though the graphical method is easy to understand, it can only be used when the number of variables in a linear problem is limited to two. Simplex methods are found to be very efficient in solving LP problems of any magnitude. The simplex method begins with a basic feasible solution and at each iteration it projects an improved solution over the earlier step. A solution is considered optimum when there is no further improvement. Simplex methods require formulation of the LP in a standard form in which all the inequality constraints are converted into equality constraints by using slack, surplus, or artificial variables. Then key operations at each iteration are performed to obtain the optimal solution. Published works on application of LP in textile manufacturing are also illustrated in this chapter.

References

Dantzig, G. B. 1963. *Linear Programming and Extensions*, Princeton University Press, Princeton, NJ.

Deb, K. 2010. *Optimization for Engineering Design*, PHI Learning, New Delhi, India.

Gupta, D., Garg, N., Arora, K. and Priyadarshini, N. 2006. Developing body measurement charts for garment manufacture based on a linear programming approach. *Journal of Textile and Apparel, Technology and Management*, 5(1), Winter.

Harianto, R. A. 2018. Optimization of woven fabric production in textile industry of PT. Argo Pantes Tangerang. *International Journal of Advanced Scientific Research and Development*, 5(4/I), 70–76.

Kang, T. J. and Lee, J. K. 1988. Application of linear programming to wool blending in the Korean textile mills, Advanced Workshop: "Application of mathematics and physics in the wool industry." *WRONZ and the Textile Institute*, Manchester, United Kingdom, February 9–12, 43–51.

LaFerney, P. E. 1969. Cotton quality, price and use value: A statistical model of a textile processing plant. *Southern Journal of Agriculture Economics*, December, 91–97.

Lee, J. K. and Kin, Y. S. 1984. *Linear Programming for Cotton Mixing: Computer in the World of Textiles*, The Textile Institute, Manchester, United Kingdom. 90.

Majumdar, A., Singh, S. P. and Ghosh, A. 2011. Modelling, optimization and decision making techniques in designing of functional clothing. *Indian Journal of Fibre and Textile Research*, 36, 398–409.

Murthy, K. 1983. *Linear Programming*, Wiley, New York.

Panneerselvam, R. 2002. *Operations Research*, Prentice Hall, New Delhi, India.

Raju, N. V. S. 2009. *Operations Research*, SMS Education, New Delhi, India.

Sharma, J. K. 2010. *Operations Research, Theory and Applications*, Macmillan, New Delhi, India.

Srinivasan, R. 2014. *Strategic Business Decisions: A Quantitative Approach*, Springer, New Delhi, India.

Taha, H. A. 1989. *Operations Research*, Macmillan, New York.

Tesfaye, G., Berhane, T., Zenebe, B. and Asmelash, S. 2016. A linear programming method to enhance resource utilization case of Ethiopian apparel sector. *International Journal for Quality Research*, 10(2), 421–432.

Vohra, N. D. 2010. *Quantitative Techniques in Management*, Tata McGraw-Hill, New Delhi, India.

7

Fuzzy Linear Programming

7.1 Introduction

In Chapter 6, linear programming (LP) problem methods are discussed, which are found to be capable of providing optimal solutions to many optimization problems where both objective and constraint functions are linear in nature. The LP problem is unarguably the most simple and commonly used classical optimization technique that has been used to solve many engineering optimization problems (Deb, 2010). However, most of the engineering problems involve parameters that are linear yet linguistic in nature and are not specified in precise or crisp terms. Such problems can be dealt with using fuzzy linear programming (FLP).

The word *fuzzy* means "vagueness" or "ambiguity." In the real world, problems often turn out to be complex due to uncertainty that arises mainly due to partial information about the problems, inherent ambiguity in the language, conflicting information from multiple sources, or information that is not fully trustworthy. In such cases, *fuzzy set theory* is an excellent tool to handle such uncertainty arising due to vagueness or ambiguity. The concept of fuzzy logic was propounded by Lotfi A. Zadeh (1965). Since then, a lot of theoretical developments have taken place, which are explained in many standard books authored by Zimmerman (1996), Cox (1998), Hung and Walker (1999), Berkan and Trubatch (2000), and Kartalopoulos (2000). To understand an FLP algorithm, the reader should initially comprehend the concept of crisp set, fuzzy set, and membership function.

7.2 Crisp Set, Fuzzy Set, and Membership Function

A classical crisp set is a container that wholly includes or wholly excludes any given element. For example, if we consider the query, "Is the fabric greige or bleached?" the reply would be a definite "yes" or "no." If "yes" and "no" are accorded values of 1 and 0, respectively, the outcome is binary (0/1) and is termed *crisp* in the domain of fuzzy set theory.

Suppose that we have a crisp set A that contains the individual elements x. Then, mathematically, we can write

$$\mu_A(x) = \begin{cases} 1, & \text{if } x \in A \\ 0, & \text{if } x \notin A \end{cases} \qquad (7.1)$$

where $\mu_A(x)$ indicates the unambiguous membership of element x in set A; $\mu_A(x)$ is either 0 or 1.

The fuzzy set includes elements with only a partial degree of "membership of acceptance or belongingness." Membership of an element in the fuzzy set is not a matter of affirmation or rejection, but merely its degree of belongingness. For example, if we consider the statement, "Is the yarn fine?" the response would not be a definite "yes" or "no"; instead, a variety of answers spanning a range, such as, if the yarn is very fine, then it may be said fine with a membership value of 1; if it is medium fine, then it may be considered as fine with a membership value of 0.6; if it is coarse, then it may be assumed to be fine with a membership value of 0.1; if it is very coarse, then also it is called fine with a membership value of 0; and so on. The circumstance in this example is that we have values between 0 and 1 in contrast to the crisp set, which has values either 0 or 1. Such a situation is known as fuzzy.

A crisp and fuzzy set is illustrated in Figure 7.1. The fundamental difference between the crisp and fuzzy set lies in the nature of membership function, which is illustrated in Figure 7.2. For a crisp set, the membership functions of its elements are binary in nature—that is, each element has a value of either 0 or 1; whereas for a fuzzy set, the membership function maps its elements onto a space in the interval of 0–1. Symbolically, the membership function of an element in a fuzzy set can be expressed as

$$\mu_{\tilde{A}}(x) \in [0,1] \qquad (7.2)$$

where $\mu_{\tilde{A}}(x)$ is the membership function of element x in fuzzy set $\tilde{A}$. Commonly, fuzzy set $\tilde{A}$ is expressed in terms of ordered pairs as

$$\tilde{A} = \{x, \mu_{\tilde{A}}(x) | x \in X\} \qquad (7.3)$$

All properties of a crisp set are applicable for a fuzzy set except the excluded-middle laws. In crisp set theory, the union of a crisp set with its complement yields the universe, and the intersection of a crisp set with its complement is null; whereas in fuzzy set theory, the union and intersection of a fuzzy set with its complement do not yield the universe and null, respectively. The difference is as follows:

$$\text{For crisp sets,} \quad \begin{matrix} A \cup A^C = X \\ A \cap A^C = \varnothing \end{matrix} \qquad (7.4)$$

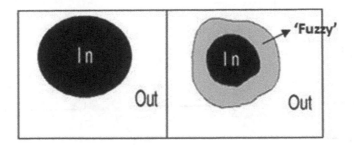

FIGURE 7.1
Crisp set (left) and fuzzy set (right).

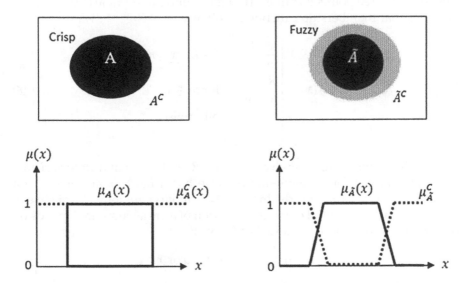

FIGURE 7.2
Representation of membership functions for crisp and fuzzy sets.

$$\text{For fuzzy sets,} \quad \begin{aligned} \tilde{A} \cup \tilde{A}^C &\neq X \\ \tilde{A} \cap \tilde{A}^C &\neq \varnothing \end{aligned} \quad (7.5)$$

7.2.1 Fuzzy Set Operations

Once the fuzzy sets are chosen, a membership function for each set should be allotted. The process of assigning membership functions to the sets of data is referred to as *fuzzification*. In this process, the crisp values are converted into fuzzy values to express uncertainties present in the crisp values. The membership function can have various forms, such as triangle, trapezoid, Gaussian, etc., which are illustrated in Figure 7.3.

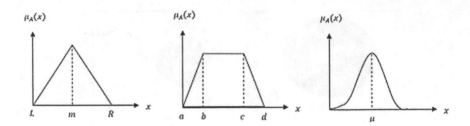

FIGURE 7.3
Various forms of membership functions.

The triangular membership function is the simplest one, and it is a collection of three points forming a triangle. Dubois and Prade (1979) defined the triangular membership function as follows:

$$\mu_{\tilde{A}}(x) = \begin{cases} \dfrac{x-L}{m-L}, & \text{for } L \leq x \leq m \\ \dfrac{R-x}{R-m} & \text{for } m \leq x \leq R \\ 0, & \text{otherwise} \end{cases} \tag{7.6}$$

where m is the most promising value, L and R are the left and right spread.

The trapezoidal form membership curve has a flat top, and it is just a truncated triangle curve producing $\mu_{\tilde{A}}(x) = 1$ in large regions of universe of discourse. The trapezoidal curve is a function of a variable x and depends on four scalar parameters $a, b, c,$ and d as follows:

$$\mu_{\tilde{A}}(x) = \begin{cases} 0, & \text{for } x \leq a \text{ or } x \geq d \\ \dfrac{x-a}{b-a} & \text{for } a \leq x \leq b \\ 1, & \text{for } b \leq x \leq c \\ \dfrac{d-x}{d-c}, & \text{for } c \leq x \leq d \end{cases} \tag{7.7}$$

The Gaussian form membership function depends on two parameters, namely, standard deviation (σ) and mean (μ), and it is represented as follows:

$$\mu_{\tilde{A}}(x) = e^{\frac{-(x-\mu)^2}{2\sigma^2}} \tag{7.8}$$

In fuzzy set operations, the membership function of the "union" of two fuzzy sets is the maximum of the two individual membership functions and is known as the *maximum criterion*. Again, the membership function of the "intersection" of two fuzzy sets is the minimum of the two individual membership functions and is known as the *minimum criterion*. Further, the membership function of the complement of a fuzzy set is the negation of the specified membership function and is known as the *negation criterion*. Symbolically, for the set $\tilde{D} = \tilde{A} \cup \tilde{B}$ ("union" of fuzzy sets $\tilde{A}$ and $\tilde{B}$), the membership function is defined as

$$\mu_{\tilde{D}}(x) = \mu_{\tilde{A}}(x) \vee \mu_{\tilde{B}}(x) = \max\{\mu_{\tilde{A}}(x), \mu_{\tilde{B}}(x)\}, x \in X \tag{7.9}$$

where the symbol $\vee$ stands for the maximum operator.

For the set $\tilde{E} = \tilde{A} \cap \tilde{B}$ ("intersections" of fuzzy sets $\tilde{A}$ and $\tilde{B}$), the membership function is defined as

$$\mu_{\tilde{E}}(x) = \mu_{\tilde{A}}(x) \wedge \mu_{\tilde{B}}(x) = \min\{\mu_{\tilde{A}}(x), \mu_{\tilde{B}}(x)\}, x \in X \tag{7.10}$$

where the symbol $\wedge$ represents the minimum operator.

The membership function of the complement of a fuzzy set $\tilde{A}$ is defined by

$$\mu_{\tilde{A}^c}(x) = \{1 - \mu_{\tilde{A}}(x)\}, x \in X \tag{7.11}$$

In more general terms, fuzzy union is defined by the fuzzy OR operator, intersection is defined by the fuzzy AND operator, and complement is defined by the fuzzy NOT operator.

7.3 Fuzzy Linear Programming Algorithm

There is no doubt that LP is not sufficient in many practical situations in which the constraints are not specified in precise crisp terms. The FLP method could be a more reasonable approach in such circumstances (Delgado et al., 1989).

To understand FLP, it is necessary to comprehend the concept of fuzzy decision. Bellman and Zadeh (1970) defined "fuzzy decision," which is important for fuzzy optimization. If $\tilde{A}$ and $\tilde{B}$ are two fuzzy sets, then the intersection of $\tilde{A}$ and $\tilde{B}$ defines the fuzzy decision. The intersection of these two fuzzy sets can be expressed by their membership functions using Equation 7.10. Thus, symbolically we can write

$$\tilde{E} = \tilde{A} \cap \tilde{B}$$

$$\therefore \mu_{\tilde{E}}(x) = \lambda = \mu_{\tilde{A}}(x) \wedge \mu_{\tilde{B}}(x) = \min\{\mu_{\tilde{A}}(x), \mu_{\tilde{B}}(x)\}, x \in X \tag{7.12}$$

where the symbol $\wedge$ represents the minimum operator.

In Equation 7.12, the minimum value of the membership function of the element x in the fuzzy set $\tilde{A}$ and fuzzy set $\tilde{B}$ defines the membership function for fuzzy set $\tilde{E}$. Now, within this membership function for fuzzy set $\tilde{E}$, the value of x (denoted by $\hat{x}$) that maximizes the membership function of fuzzy set $\tilde{E}$ (denoted by $\hat{\lambda}$) is the optimum value. Therefore, we can write

$$\mu_{\tilde{E}}(\hat{x}) = \hat{\lambda} = \max[\mu_{\tilde{E}}(x)], x \in X \tag{7.13}$$

The concept of fuzzy optimization of the above-mentioned two fuzzy sets ($\tilde{A}$ and $\tilde{B}$) is illustrated in Figure 7.4. The membership functions for the two goals (i.e., either objective and constraint; or, two constraints) that are assumed as linear are shown in Figure 7.4.

From the figure, it is obvious that for the membership function of the goal $\tilde{A}$, as x increases the acceptability decreases, whereas the trend is reversed for the membership function of the other goal. Thus, there lies a conflicting statement between the two goals, which are defined by their corresponding membership functions. Therefore, the intersection of the minimum values between the membership functions of goals $\tilde{A}$ and $\tilde{B}$ is the membership function of the fuzzy decision set $[\mu_{\tilde{E}}(x) = \lambda = \min\{\mu_{\tilde{A}}(x), \mu_{\tilde{B}}(x)\}]$ and is denoted by the darker continuous line in Figure 7.4. Once the membership function of the fuzzy decision set is defined, x that maximizes the membership function of the fuzzy decision set (λ) is calculated. Then $\hat{\lambda}$ is the maximum membership value of the membership function of the fuzzy decision, and $\hat{x}$ is its associated value. So, $\hat{x}$ indicates the best compromised solution and gives the optimum value. In general, there are large numbers of fuzzy constraints or a combination of fuzzy and crisp constraints.

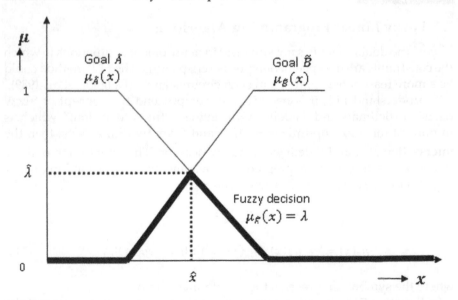

FIGURE 7.4
Concept of fuzzy optimization.

The general type of FLP can be formulated as follows (Klir and Yuan 2000):

$$\text{Maximize} \quad Z = \sum_{j=1}^{n} c_j x_j$$

subject to,

$$\sum_{j=1}^{n} a_{ij} x_j \leq B_i \quad (i \in \mathbb{N}^m)$$

$$x_j \geq 0 \quad (j \in \mathbb{N}^n)$$

(7.14)

where $\mathbb{N}^m$ is the collection of m-tuples over natural number $\mathbb{N}$, and it is expressed as $\mathbb{N}^m = \{x_1, x_2, \cdots, x_m \,|\, x_i \in \mathbb{N}\}$; $\mathbb{N}^n = \{x_1, x_2, \cdots, x_n \,|\, x_i \in \mathbb{N}\}$; $B_i (i \in \mathbb{N}^m)$ is the fuzzy number of the ith constraint having the following form:

$$B_i(x) = \begin{cases} 1 & \text{when } x \leq b_i \\ \dfrac{b_i + p_i - x}{p_i} & \text{when } b_i < x < b_i + p_i \\ 0 & \text{when } b_i + p_i \leq x \end{cases}$$

(7.15)

where $x \in \mathbb{R}$, $\mathbb{R}$ is real number space, and p_i is the maximum tolerance from b_i, as determined by the decision maker (refer to Figure 7.5).

Now, for each vector $x = (x_1, x_2, \ldots x_n)$, the degree, $D_i(x)$, to which x satisfies the ith constraint $(i \in \mathbb{N}^m)$ is given by

$$D_i(x) = B_i \left(\sum_{j=1}^{n} a_{ij} x_j \right)$$

(7.16)

where $D_i(x)$ are the fuzzy sets on $\mathbb{R}^m$, and their intersection, $\cap_{i=1}^{m} D_i$, is a fuzzy feasible set.

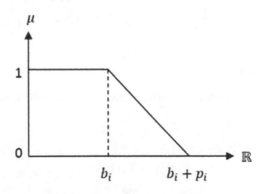

FIGURE 7.5
Fuzzy numbers employed in a general FLP problem.

In the next step, the fuzzy set of optimal values is determined by calculating the lower and upper bounds of the optimal values. The lower bound of the optimal values, Z_l, is obtained by solving the following problem:

$$\text{Maximize} \quad Z = \sum_{j=1}^{n} c_j x_j$$

subject to,

$$\sum_{j=1}^{n} a_{ij} x_j \le b_i \quad (i \in \mathbb{N}^m) \tag{7.17}$$

$$x_j \ge 0 \quad (j \in \mathbb{N}^n)$$

The upper bound of the optimal values, Z_u, is obtained by a similar problem in which each b_i is replaced with $b_i + p_i$:

$$\text{Maximize} \quad Z = \sum_{j=1}^{n} c_j x_j$$

subject to,

$$\sum_{j=1}^{n} a_{ij} x_j \le b_i + p_i \quad (i \in \mathbb{N}^m) \tag{7.18}$$

$$x_j \ge 0 \quad (j \in \mathbb{N}^n)$$

Then, the fuzzy set of optimal values, G, which is a fuzzy subset of $\mathbb{R}^n$, is defined by

$$G(x) = \begin{cases} 1 & \text{when } z_u \le cx \\ \dfrac{cx - z_l}{z_u - z_l} & \text{when } z_l < cx < z_u \\ 0 & \text{when } cx \le z_l \end{cases} \tag{7.19}$$

where $cx = \sum_{j=1}^{n} c_j x_j$.

Now the problem becomes the following classical optimization problem:

$$\text{Maximize } \lambda$$

subject to,

$$G(x) \ge \lambda \tag{7.20}$$
$$B_i(x) \ge \lambda$$
$$\lambda, x_j \ge 0 \quad (j \in \mathbb{N}^n)$$

This problem is actually a problem of finding $x \in \mathbb{R}^n$ such that $\left[\left(\cap_{i=1}^m D_i\right) \cap G\right](x)$ reaches the maximum value—that is, a problem of finding a point that satisfies the constraints and goal with the maximum degree.

EXAMPLE 7.1

Let us consider an export house makes two different styles of apparel (say *Style P* and *Style Q*). The profits of *Style P* and *Style Q* are Rs. 40 and Rs. 30, respectively. Each of *Style P* and *Style Q* requires 1 meter of fabric. The company has total weekly availability of at least 1400 meters of fabric, but this may be extended to 1,600 meters per day. The total man-minute available in the company per week is 16,800 but can be extended to 18,000. It takes 20 and 10 man-minutes to manufacture one *Style P* and *Style Q* apparel, respectively. The manufacturer wants to know the number of *Style P* and *Style Q* to be manufactured per week to maximize profit.

The problem is stated as follows:

$$
\begin{array}{lll}
\text{Maximize the profit} & Z = 40x_1 + 30x_2 & \text{(profit)} \\
\text{Subject to} & x_1 + x_2 \leq B_1 & \text{(fabric)} \\
& 20x_1 + 10x_2 \leq B_2 & \text{(man-minutes)} \\
& x_1, x_2 \geq 0 &
\end{array}
\tag{7.21}
$$

where x_1 is the number of apparel in *Style P*, x_2 is the number of apparel in *Style Q*, B_1 and B_2 are membership functions for the constraints "fabric" and "man-minutes," respectively.

B_1 is defined by

$$
B_1(x) = \begin{cases} 1 & \text{when } (x_1 + x_2) \leq 1,400 \\ \dfrac{1,600 - (x_1 + x_2)}{200} & \text{when } 1,400 < (x_1 + x_2) < 1,600 \\ 0 & \text{when } 1,600 \leq (x_1 + x_2) \end{cases}
\tag{7.22}
$$

B_2 is defined by

$$
B_2(x) = \begin{cases} 1 & \text{when } (20x_1 + 10x_2) \leq 16,800 \\ \dfrac{18,000 - (20x_1 + 10x_2)}{1,200} & \text{when } 16,800 < (20x_1 + 10x_2) < 18,000 \\ 0 & \text{when } 18,000 \leq (20x_1 + 10x_2) \end{cases}
\tag{7.23}
$$

At first the lower (Z_l) and upper bounds (Z_u) of the objective functions are calculated from two classical linear programming problems as follows. The solutions are $Z_l = 44{,}800$ and $Z_u = 50{,}000$.

$$(P_1)\max Z = 40x_1 + 30x_2$$

$$
\begin{aligned}
\text{Subject to} \quad & x_1 + x_2 \leq 1{,}400 && \text{(fabric)} \\
& 20x_1 + 10x_2 \leq 16{,}800 && \text{(man-minutes)} \\
& x_1, x_2 \geq 0
\end{aligned}
\tag{7.24}
$$

$$(P_2)\max Z = 40x_1 + 30x_2$$

$$
\begin{aligned}
\text{Subject to} \quad & x_1 + x_2 \leq 1{,}600 && \text{(fabric)} \\
& 20x_1 + 10x_2 \leq 18{,}000 && \text{(man-minutes)} \\
& x_1, x_2 \geq 0
\end{aligned}
\tag{7.25}
$$

The membership function of the optimal values G, is defined by

$$G(x) = \frac{(40x_1 + 30x_2) - 44{,}800}{50{,}000 - 44{,}800} \tag{7.26}$$

The membership function plots for different constraints are illustrated in Figure 7.6.

Consequently, the FLP becomes

$$
\begin{aligned}
\text{Maximize} \quad & \lambda \\
\text{Subject to} \quad & \frac{(40x_1 + 30x_2) - 44{,}800}{50{,}000 - 44{,}800} \geq \lambda \\
& \text{or,} - 40x_1 - 30x_2 + 5{,}200\lambda \leq -44{,}800 \\
& \frac{1{,}600 - (x_1 + x_2)}{200} \geq \lambda \\
& \text{or,}\ x_1 + x_2 + 200\lambda \leq 1{,}600 \\
& \frac{18{,}000 - (20x_1 + 10x_2)}{1{,}200} \geq \lambda \\
& \text{or,}\ 20x_1 + 10x_2 + 1{,}200\lambda \leq 18{,}000 \\
& \text{and,}\ x_1, x_2, \lambda \geq 0
\end{aligned}
\tag{7.27}
$$

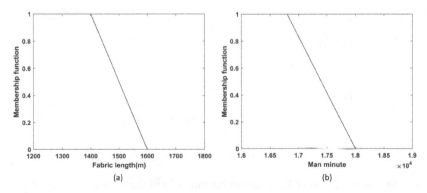

(a)   (b)

FIGURE 7.6
Membership function plots of (a) fabric and (b) man-minute.

Solving this classical optimization problem, we get $\hat{\lambda}$ (i.e., maximum λ) = 0.5 for $\hat{x}_1 = 240$ and $\hat{x}_2 = 1,260$. Therefore, the maximum profit, $\hat{Z} = 40\hat{x}_1 + 30\hat{x}_2 = $ Rs. $47,400$.

The MATLAB® coding for the aforesaid optimization problem using FLP is given in Section 7.5.

EXAMPLE 7.2

A spinning mill plans to spin 50s Ne 100% cotton combed yarn from two different fibers (MCU5 and S6). The mill wants to determine the proportion of MCU5 and S6 cotton fibers to achieve the best-quality yarn at the lowest material cost. Three fiber parameters were considered, namely, fiber strength (cN/tex), upper-half mean length (UHML), and fiber fineness. The fiber properties and their costs are shown in Table 7.1. The minimum fiber strength, upper-half mean length, and fiber fineness to spin 50s Ne with desired quality are 31.5 cN/tex, 30 mm, and 4 micronaire, respectively. However, from past experience, it has been decided that a compromise can be made in all of these fiber parameters, with maximum up to 31 cN/tex, 29.5 mm, and 4.1 micronaire, with a lesser degree of acceptance.

Therefore, the problem is stated as follows:

$$
\begin{aligned}
\text{Minimize} \quad & Z = 130x_1 + 120x_2 \quad \text{(mixing cost)} \\
\text{Subject to} \quad & 34x_1 + 30x_2 \geq B_1 \quad \text{(fiber strength)} \\
& 34x_1 + 28x_2 \geq B_2 \quad \text{(UHML)} \\
& 3.7x_1 + 4.3x_2 \leq B_3 \quad \text{(fiber fineness)} \\
& x_1 + x_2 = 1 \\
& x_1, x_2 \geq 0
\end{aligned}
\tag{7.28}
$$

where x_1 is the proportion of MCU5 cotton fiber; x_2 is the proportion of S6 cotton fiber; B_1, B_2, and B_3 represent the membership functions for the constraints of fiber strength, UHML, and fiber fineness, respectively.

The membership functions for the constraints of fiber strength, UHML, and fiber fineness are decided by the spinning mill and accordingly defined as follows:

TABLE 7.1

Fiber Properties and Cost of MCU5 and Bunny

Cotton	Fiber Strength (cN/tex)	UHML (mm)	Fiber Fineness (µg/inch)	Cost/kg (Rs./kg)
MCU5	34	34	3.7	130
S 6	30	28	4.3	120

B_1 is defined by

$$B_1(x) = \begin{cases} 1 & \text{when } (34x_1 + 30x_2) \geq 31.5 \\ \dfrac{(34x_1 + 30x_2) - 31}{31.5 - 31} & \text{when } 31 < (34x_1 + 30x_2) < 31.5 \\ 0 & \text{when } 31 \geq (34x_1 + 30x_2) \end{cases} \tag{7.29}$$

B_2 is defined by

$$B_2(x) = \begin{cases} 1 & \text{when } (34x_1 + 28x_2) \geq 30 \\ \dfrac{(34x_1 + 28x_2) - 29.5}{30 - 29.5} & \text{when } 29.5 < (34x_1 + 28x_2) < 30 \\ 0 & \text{when } 29.5 \geq (34x_1 + 28x_2) \end{cases} \tag{7.30}$$

B_3 is defined by

$$B_3(x) = \begin{cases} 1 & \text{when } (3.7x_1 + 4.3x_2) \leq 4 \\ \dfrac{4.1 - (3.7x_1 + 4.3x_2)}{4.1 - 4} & \text{when } 4 < (3.7x_1 + 4.3x_2) < 4.1 \\ 0 & \text{when } 4.1 \leq (3.7x_1 + 4.3x_2) \end{cases} \tag{7.31}$$

At first, the lower (Z_l) and upper bounds (Z_u) of the objective functions are calculated from two classical linear programming problems as follows. The solutions are $Z_u = 125$ and $Z_l = 123.33$.

$$
\begin{aligned}
&(P_1) \text{ Minimize} && Z = 130x_1 + 120x_2 \\
&\text{Subject to} && 34x_1 + 30x_2 \geq 31.5 \\
& && 34x_1 + 28x_2 \geq 30 \\
& && 3.7x_1 + 4.3x_2 \leq 4 \\
& && x_1 + x_2 = 1 \\
& && x_1, x_2 \geq 0
\end{aligned} \tag{7.32}
$$

$$
\begin{aligned}
&(P_2) \text{ Minimize} && Z = 130x_1 + 120x_2 \\
&\text{Subject to} && 34x_1 + 30x_2 \geq 31 \\
& && 34x_1 + 28x_2 \geq 29.5 \\
& && 3.7x_1 + 4.3x_2 \leq 4.1 \\
& && x_1 + x_2 = 1 \\
& && \text{and } x_1, x_2 \geq 0
\end{aligned} \tag{7.33}
$$

The membership function of the optimal values G is defined by

$$G(x) = \frac{(130x_1 + 120x_2) - 123.33}{125 - 123.33} \tag{7.34}$$

The membership function plots for different constraints are illustrated in Figure 7.7.

Consequently, the FLP becomes

Maximize λ

Subject to $\dfrac{(130x_1 + 120x_2) - 123.33}{125 - 123.33} \geq \lambda$

or, $130x_1 + 120x_2 + 1.67\lambda \leq 125$

$\dfrac{(34x_1 + 30x_2) - 31}{31.5 - 31} \geq \lambda$

or, $34x_1 + 30x_2 - 0.5\lambda \geq 31$

$\dfrac{(34x_1 + 28x_2) - 29.5}{30 - 29.5} \geq \lambda$ (7.35)

or, $34x_1 + 28x_2 - 0.5\lambda \geq 29.5$

$\dfrac{4.1 - (3.7x_1 + 4.3x_2)}{4.1 - 4} \geq \lambda$

or, $3.7x_1 + 4.3x_2 + 0.1\lambda \leq 4.1$

and $x_1, x_2, \lambda \geq 0$

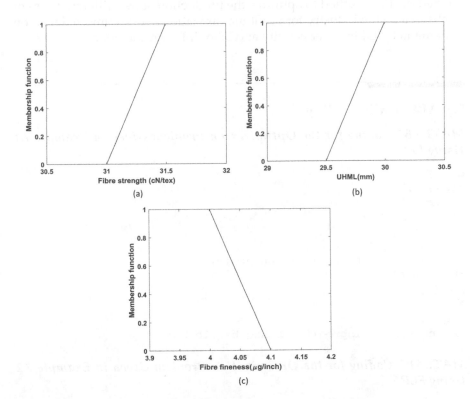

FIGURE 7.7
Membership function plots of (a) fiber strength, (b) UHML, and (c) fiber fineness.

Solving the above classical optimization problem, we get $\hat{\lambda}$ (i.e., maximum λ) $=0.5$ for $\hat{x}_1 = 0.42$ and $\hat{x}_2 = 0.58$.

Therefore, the minimum mixing cost $\hat{Z} = 130\hat{x}_1 + 120\hat{x}_2 =$ Rs. 124.2.

The MATLAB® coding for the aforesaid optimization problem using FLP is presented in Section 7.5.

7.4 Applications

Few works have been reported on the application of FLP in the textile industry. Elamvazuthi et al. (2009) have tried to determine the monthly production planning quotas and profits of a home textile industry using an FLP problem. Das and Ghosh (2015) utilized the FLP method to minimize the costs of three different types of cotton mixings suitable for the production of fine, medium, and coarse yarns without impacting the desired quality. They found that the FLP attains the best solution in comparison to classical linear programming and the existing method followed by the textile industry. Teke et al. (2017) applied the FLP method to optimize the production of five different types of cloths in a textile industry for profit maximization and minimum deviation of demand by taking productivity and labor into consideration.

7.5 MATLAB® Coding

MATLAB® Coding for the Optimization Problem Given in Example 7.1 Using FLP

```
clc
close all
clear all
format short g
f=-[1 0 0];
b=[-44800 1600 18000];
A=[5200 -40 -30;200 1 1;1200 20 10];
lb=[0 0 0];
ub=[];
Aeq=[];
beq=[];
[x fn_value]=linprog(f,A,b,Aeq,beq,lb,ub)
```

MATLAB® Coding for the Optimization Problem Given in Example 7.2 Using FLP

```
clc
close all
```

```
clear all
format short g
f=-[1 0 0];
b=[125 -31 -29.5 4.1];
A=[1.67 130 120;0.5 -34 -30;0.5 -34 -28; 0.1 3.7 4.3];
lb=[0 0 0];
ub=[];
Aeq=[0 1 1];
beq=[1];
[x fn_value]=linprog(f,A,b,Aeq,beq,lb,ub)
```

7.6 Summary

The LP problem method is unarguably the most simple and commonly used classical optimization technique that has been used to solve many engineering optimization problems when both the objective and constraints functions are linear. However, most of the practical problems in engineering manufacturing are linguistic in nature; hence, they are not specified in precise or crisp terms. Such problems are dealt with in the FLP method. In this chapter, the fundamentals of crisp sets and fuzzy sets are explained. The membership function is defined, which is associated with the fuzzy set. Different fuzzy operations are explained. Then the FLP problem-solving algorithm is demonstrated step by step. Two examples of the FLP problem are cited and discussed step by step. MATLAB® coding for these examples is given. Last, the applications of FLP in the textile industry are highlighted.

References

Bellman, R. E. and Zadeh, L. A. 1970. Decision making in a fuzzy environment. *Management Science*, 17, 141–164.

Berkan. R. C. and Trubatch, S. L. 2000. *Fuzzy Systems Design Principles*, Standard Publishers Distributers, New Delhi, India.

Cox, E. 1998. *The Fuzzy Systems Handbook*, Morgan Kaufmann, San Francisco, CA.

Das, S. and Ghosh, A. 2015. Optimization of cotton mixing cost: A fuzzy linear programming approach. *ACC Journal*, 21(1), 24–29.

Deb, K. 2010. *Optimization for Engineering Design*, New Delhi, India: PHI Learning.

Delgado, M., Verdegay, J. L. and Vila, M. A. 1989. A general model for fuzzy linear programming. *Fuzzy Sets and Systems*, 29, 21–30.

Dubois, D. and Prade, H. 1979. Fuzzy real algebra, some results. *Fuzzy Sets and Systems*, 2, 327–348.

Elamvazuthi, I., Ganesan, T. and Vasant, P. 2009. Application of a fuzzy programming technique to production planning in the textile industry. *International Journal of Computer Science and Information Security*, 6(3), 238–243.

Hung, T. N. and Walker, E. A. 1999. *A First Course in Fuzzy Logic*, CRC Press, Boca Raton, FL.

Kartalopoulos, S. V. 2000. *Understanding Neural Networks and Fuzzy Logic*, Prentice Hall, New Delhi, India.

Klir, G. J. and Yuan, B. 2000. *Fuzzy Sets and Fuzzy Logic: Theory of Applications*, Prentice Hall, New Delhi, India.

Teke, C., Okuntan, C. and Erden, C. 2017. Determining the production amounts in textile industry with fuzzy linear programming. *International Journal of Engineering and Technology Research*, 2, 1–6.

Zadeh, L. A. 1965. Fuzzy sets. *Information and Control*, 8, 338–353.

Zimmerman, H. J. 1996. *Fuzzy Set Theory and Its Applications*, 2nd ed., Allied, New Delhi, India.

8

Quadratic Programming

8.1 Introduction

The linear programming (LP) technique is a powerful model for solving various types of optimization problems in the manufacturing process. But, the LP method has its limitations (Sharma, 2010). In the textile manufacturing process, all phenomena are not linear. Hence, the LP problem model cannot be used when the response and constraints functions are nonlinear in nature (quadratic, hyperbolic, exponential, etc.). When the complexity of the problem goes to the next level, it becomes non-linear problem (the non-linear problem has finite number of decision variables), subject to finite number of constraint functions (These finite constraint functions are linear equality and/or non-equality constraint functions.), the nonlinear programming process model can be used to solve the optimization problem. Among several nonlinear programming processes that are used to solve real-life problems faced during manufacturing, the quadratic programming process is used when the problem has a nonlinear quadratic objective function and linear constraint functions (Panneerselvam, 2002). In this chapter, quadratic programming with linear constraints is discussed.

8.2 Quadratic Programming Algorithm

The general structure of a quadratic programming problem (QPP) consists of a quadratic objective function and linear constraint functions as follows:

$$\text{Optimize (max or min)} \quad Z = \left\{ \sum_{j}^{n} c_j x_j + \frac{1}{2} \sum_{j=1}^{n} \sum_{k=1}^{n} x_j d_{jk} x_k \right\}$$

subject to constraints

$$\sum_{j=1}^{n} a_{ij} x_j \leq b_i$$

and $\quad x_j \geq 0$ for all j

(8.1)

where $i = 1, 2, 3, \ldots, m; j = 1, 2, 3, \ldots, n; d_{jk} = d_{kj}; b_i \geq 0$ for all $i = 1, 2, 3\ldots, m$.

In order to understand how to solve a QPP, the reader should know Lagrangian function and Kuhn-Tucker conditions.

8.2.1 Lagrangian Function

A *Lagrangian function* is a technique that combines both the objective and constraint functions of a problem into a single equation by introducing a multiplier known as the *Lagrangian multiplier*. The Lagrangian multiplier measures the change that occurs in the objective function per unit change in the constraint. Solving the Lagrangian function allows us to solve the given optimization problem (Deb, 2010). A simple example of formation of the Lagrangian function follows.

Assume an objective function Z is being optimized, which is a function of x_1 and x_2. The constraint functions are $g_1(x_1, x_2) \leq 0$ and $g_2(x_1, x_2) \leq 0$. Therefore, the problem can be written as

$$
\begin{aligned}
\text{Optimize} \quad & Z = f(x_1, x_2) \\
\text{subject to} \quad & g_1(x_1, x_2) \leq 0 \\
& g_2(x_1, x_2) \leq 0 \\
\text{and } & x_1, x_2 \geq 0
\end{aligned}
\tag{8.2}
$$

These inequality constraints are converted into equations by introducing slack variables s_1^2 and s_2^2. (s_1^2 and s_2^2 are used to ensure the nonnegative value of the slack variable is added.) Therefore, the constraint functions can be written as

$$
g_1(x_1, x_2) + s_1^2 = 0
$$

$$
g_2(x_1, x_2) + s_2^2 = 0
$$

The Lagrangian function (L) is formulated combining both the objective function, $Z = f(x_1, x_2)$, and constraint functions $g_1(x_1, x_2) + s_1^2 = 0$ and $g_2(x_1, x_2) + s_2^2 = 0$ by introducing multipliers (λ_1 and λ_2), known as Lagrangian multipliers. The Lagrangian function is given as follows:

$$
L(x_1, x_2, s_1, s_2, \lambda_1, \lambda_2) = f(x_1, x_2) - \lambda_1 \left\{ g_1(x_1, x_2) + s_1^2 \right\} - \lambda_2 \left\{ g_2(x_1, x_2) + s_2^2 \right\}
\tag{8.3}
$$

Therefore, the generalized form of a Lagrangian equation can be expressed as follows:

$$
L(x, s, \lambda) = f(x) - \sum_{i=1}^{m} \lambda_i \left[g_i(x) + s_i^2 \right]
\tag{8.4}
$$

where $f(x)$ is the objective function; $g_i(x) \leq 0$ are the constraints; s_i^2 are the slack variables; λ_i are the Lagrangian multipliers; and $i = 1, 2, 3, \dots, m$.

8.2.2 Kuhn-Tucker Conditions

Kuhn and Tucker (1951) obtained necessary conditions for a local optimum and, in some cases, necessary and sufficient conditions for the global optimum for nonlinear programming problems. Wolfe has applied the Kuhn-Tucker theory to the QPP to obtain optimal solutions.

Consider the following objective function:

$$\begin{aligned} \text{Maximize} \quad & Z = f(x) \\ \text{subject to} \quad & h_i(x) \leq b_i \\ & \text{and } x \geq 0 \end{aligned} \tag{8.5}$$

where $x = (x_1, x_2, x_3, \dots, x_n)$ and $i = 1, 2, 3, \dots, m$.

The inequality constraints are converted into equality constraints by adding slack variables s_i^2 to each constraint function. (s_i^2 has been used to ensure the nonnegative value of the slack variable is added.) The optimization problem can be rewritten as

$$\begin{aligned} \text{Maximize} \quad & Z = f(x) \\ \text{subject to} \quad & h_i(x) - b_i + s_i^2 = 0 \quad \text{i.e., } g_i(x) + s_i^2 = 0 \\ & \text{and } x \geq 0 \end{aligned} \tag{8.6}$$

The Lagrangian function for this optimization problem is as follows:

$$L(x, s, \lambda) = f(x) - \sum_{i=1}^{m} \lambda_i \left[g_i(x) + s_i^2 \right] \tag{8.7}$$

where λ_i is the Lagrangian multiplier; and $i = 1, 2, 3, \dots, m$.

The necessary conditions for a point to be optimal can be obtained by solving the following equations:

$$\frac{\partial L}{\partial x_j} = 0$$

$$\text{i.e., } \frac{\partial f(x)}{\partial x_j} - \sum_{i=1}^{m} \lambda_i \frac{\partial g_i(x)}{\partial x_j} = 0 \quad j = 1, 2, 3, \dots, n \tag{8.8}$$

$$\frac{\partial L}{\partial \lambda_i} = 0$$

$$\text{i.e., } -\left[g_i(x) + s_i^2 \right] = 0 \quad i = 1, 2, 3, \dots, m \tag{8.9}$$

$$\frac{\partial L}{\partial s_i} = 0$$

(8.10)

i.e., $-2\lambda_i s_i = 0$ $i = 1, 2, 3, \ldots, n$

1. Now, from the equation $\partial L / \partial s_i = 0$, we get $-2\lambda_i s_i = 0$, that is, $\lambda_i s_i = 0$.

   The equation $\lambda_i s_i = 0$ means that either $\lambda_i = 0$ or $s_i = 0$. If $s_i = 0$ and $\lambda_i > 0$, then from the equation $\partial L / \partial \lambda_i = 0$, we get $g_i(x) = 0$. Therefore, replacing s_i with $g_i(x)$ in equation $\lambda_i s_i = 0$, we get $\lambda_i g_i(x) = 0$.

2. Now, from the equation $-\left[g_i(x) + s_i^2\right] = 0$, we get $g_i(x) + s_i^2 = 0$.

   The slack variable $s_i^2 \geq 0$. It is obvious from the equation $g_i(x) + s_i^2 = 0$ that when $s_i^2 = 0$, $g_i(x) = 0$, and when $s_i^2 > 0$, $g_i(x) \leq 0$.

3. The following are the possibilities of λ_i:

   a. For the maximization objective function and "less than or equal to" type constraints ($\leq$), $\lambda_i \geq 0$.

   b. For the maximization objective function and "greater than or equal to" type constraints ($\geq$), $\lambda_i \leq 0$.

   c. For the minimization objective function and "less than or equal to" type constraints ($\leq$), $\lambda_i \leq 0$.

   d. For the minimization objective function and "greater than or equal to" type constraints ($\geq$), $\lambda_i \geq 0$.

   e. For both maximization and minimization objective functions with "equal to" type constraints ($=$), λ_i is unrestricted in sign.

The Kuhn-Tucker necessary conditions at a local optimal point for the above maximization problem are summarized as follows:

a.
$$\frac{\partial f(x)}{\partial x_j} - \sum_{i=1}^{m} \lambda_i \frac{\partial g_i(x)}{\partial x_j} = 0$$

(8.11)

b.
$$\lambda_i g_i(x) = 0$$

(8.12)

c.
$$g_i(x) \leq 0$$

(8.13)

d.
$$\lambda_i \geq 0$$

(8.14)

In constrained optimization problems, the optimal point is one that satisfies Kuhn-Tucker conditions.

8.2.3 Wolfe's Method to Solve Quadratic Programming Problem

Optimize (max or min) $\quad Z = f(x) = \left\{ \sum_{j}^{n} c_j x_j + \frac{1}{2} \sum_{j=1}^{n} \sum_{k=1}^{n} x_j d_{jk} x_k \right\}$

subject to constraints $\quad \sum_{j=1}^{n} a_{ij} x_j \leq b_i$

$$(8.15)$$

and $\quad x \geq 0$ for all i and j

where $i = 1, 2, 3, \ldots, m; j = 1, 2, 3, \ldots, n; c_{jk} = c_{kj}; b_i \geq 0$ for all $i = 1, 2, 3, \ldots, m$.

Step 1

The inequality constraints are converted into equations by introducing slack variables $s_i^2 (i = 1, 2, 3, \ldots, m)$ and $r_j^2 (j = 1, 2, 3, \ldots, n)$. Thus, the problem becomes

Optimize (max or min) $\quad Z = \left\{ \sum_{j}^{n} c_j x_j + \frac{1}{2} \sum_{j=1}^{n} \sum_{k=1}^{n} x_j d_{jk} x_k \right\}$

subject to constraints

$$(8.16)$$

$$\sum_{j=1}^{n} a_{ij} x_j + s_i^2 = b_i$$

and $-x_j + r_j^2 = 0$ for all i and j

where $i = 1, 2, 3, \ldots, m; j = 1, 2, 3, \ldots, n; c_{jk} = c_{kj}; b_i \geq 0$ for all $i = 1, 2, 3, \ldots, m$.

Step 2

The Lagrangian function (L) is formulated combining both the objective function and constraint functions by introducing the Lagrangian multiplier. The Lagrangian function is as follows:

$$L(x, s, r, \lambda, \mu) = f(x) - \sum_{i=1}^{m} \lambda_i \left[\sum_{j=1}^{n} a_{ij} x_j - b_i + s_i^2 \right] - \sum_{j=1}^{n} \mu_j \left[-x_j + r_j^2 \right] \ldots \quad (8.17)$$

Step 3

Differentiate $L(x, s, r, \lambda, \mu)$ partially with respect to the components x, s, r, λ, and μ and equate these first-order derivatives to zero to get the required Kuhn-Tucker necessary conditions:

a.
$$\frac{\partial L}{\partial x_j} = c_j - \sum_{k=1}^{n} x_k d_{jk} - \sum_{i=1}^{m} \lambda_i a_{ij} + \mu_j = 0 \qquad (8.18)$$

b.
$$\frac{\partial L}{\partial s_i} = -2\lambda_i s_i = 0$$

or $\lambda_i s_i = 0$ (complementary slackness conditions) or

$$\lambda_i \left\{ \sum_{j=1}^{n} a_{ij} x_j - b_i \right\} = 0 \qquad (8.19)$$

c.
$$\frac{\partial L}{\partial r_j} = -2\mu_j x_j = 0$$

or $\mu_j x_j = 0$ (complementary slackness conditions) $\qquad (8.20)$

d.
$$\frac{\partial L}{\partial \lambda_i} = 0$$

or

$$\sum_{j=1}^{n} a_{ij} x_j - b_i + s_i^2 = 0$$

or

$$\sum_{j=1}^{n} a_{ij} x_j \leq b_i \qquad (8.21)$$

e.
$$\frac{\partial L}{\partial \mu_j} = 0$$

or

$$-x_j + r_j^2 = 0$$

or

$$x_j \geq 0 \qquad (8.22)$$

f.
$$\lambda_i, \mu_j \geq 0 \qquad (8.23)$$

Step 4

Introduce artificial variables R_j, where $j = 1,2,3, \ldots, n$ in the Kuhn-Tucker condition (a)—that is, in

$$c_j - \sum_{k=1}^{n} x_k d_{jk} - \sum_{i=1}^{m} \lambda_i a_{ij} + \mu_j = 0$$

we get

$$c_j - \sum_{k=1}^{n} x_k d_{jk} - \sum_{i=1}^{m} \lambda_i a_{ij} + \mu_j + R_j = 0 \qquad (8.24)$$

Step 5

Apply Phase I of the simplex method for the problem:

Minimize $\qquad Z_R = R_1 + R_2 + R_3 + \cdots + R_j$

i.e., Maximize $\quad Z_R' = -Z_R = -R_1 - R_2 - R_3 - \cdots - R_j$

Subject to $\qquad c_j - \sum_{k=1}^{n} x_k d_{jk} - \sum_{i=1}^{m} \lambda_i a_{ij} + \mu_j + R_j = 0$

$$\sum_{j=1}^{n} a_{ij} x_j - b_i + s_i^2 = 0 \text{ (where } i = 1,2,3,\ldots,m) \qquad (8.25)$$

$$\lambda_i s_i = 0; \mu_j x_j = 0$$

(complementary slackness conditions)

$x_j, \lambda_j, \mu_j, s_i, R_j \geq 0$ for all i and j

It is to be noted that while deciding for a variable to enter the basis at each iteration, the complementary slackness conditions must be satisfied. The complementary slackness conditions $\lambda_i s_i = 0$ and $\mu_j x_j = 0$ imply that both λ_i and s_i as well as μ_j and x_j cannot be the basic variables at a time in a nongenerated basic feasible solution.

Step 6

The two-phase simplex method is used to find the optimum solution of the LP problem in step 5.

Step 7

The optimum solution obtained in step 6 is the optimal solution of the QPP.

A simple example of the QPP is explained step by step for better understanding.

EXAMPLE 8.1

Solving the following QPP

$$\text{Maximize} \quad Z = 4x_1 + 6x_2 - 2x_1^2 - 2x_1x_2 - 2x_2^2$$
$$\text{subject to} \quad x_1 + 2x_2 \leq 2 \tag{8.26}$$
$$\text{and } x_1, x_2 \geq 0$$

Slack variables (s_1^2, r_1^2, and r_2^2) are added to the inequality constraints to express them as equations. Thus, the problem becomes

$$\text{Maximize} \quad Z = 4x_1 + 6x_2 - 2x_1^2 - 2x_1x_2 - 2x_2^2$$
$$\text{subject to} \quad x_1 + 2x_2 + s_1^2 = 2$$
$$-x_1 + r_1^2 = 0 \tag{8.27}$$
$$-x_2 + r_2^2 = 0$$

To obtain the Kuhn-Tucker necessary condition, we construct the Lagrange function:

$$L(x_1, x_2, \lambda_1, \mu_1, \mu_2, s_1, r_1, r_2) = 4x_1 + 6x_2 - 2x_1^2 - 2x_1x_2 - 2x_2^2 - \lambda_1$$
$$\left(x_1 + 2x_2 + s_1^2 - 2\right) - \mu_1\left(-x_1 + r_1^2\right) - \mu_2\left(-x_2 + r_2^2\right) \tag{8.28}$$

The necessary and sufficient conditions for maximization of L and hence of Z are

$$\frac{\partial L}{\partial x_1} = 4 - 4x_1 - 2x_2 - \lambda_1 + \mu_1 = 0$$

$$\frac{\partial L}{\partial x_2} = 6 - 2x_1 - 4x_2 - 2\lambda_1 + \mu_2 = 0$$

$$\frac{\partial L}{\partial \lambda_1} = x_1 + 2x_2 + s_1^2 - 2 = 0$$

$$\frac{\partial L}{\partial \mu_1} = -x_1 + r_1^2 = 0$$

$$\frac{\partial L}{\partial \mu_2} = -x_1 + r_2^2 = 0$$

$$\frac{\partial L}{\partial s_1} = -2\lambda_1 s_1 = 0 \quad \text{or, } \lambda_1 s_1 = 0$$

$$\frac{\partial L}{\partial r_1} = -2\mu_1 r_1 = 0 \quad \text{or, } \mu_1 r_1 = 0$$

$$\frac{\partial L}{\partial r_2} = -2\mu_2 r_2 = 0 \quad \text{or,} \quad \mu_2 r_2 = 0$$

Now, after introducing artificial variables R_1 and R_2 in the first two Kuhn-Tucker conditions, the LP problem becomes

$$
\begin{aligned}
\text{Minimize} \quad & Z_R = R_1 + R_2 \\
\text{i.e., Maximize} \quad & Z_R' = -Z_R = -R_1 - R_2 \\
\text{subject to} \quad & 4x_1 + 2x_2 + \lambda_1 - \mu_1 + R_1 = 4 \\
& 2x_1 + 4x_2 + 2\lambda_1 - \mu_2 + R_2 = 6 \\
& x_1 + 2x_2 + s_1^2 = 2 \\
& x_1, x_2, \lambda_1, \mu_1, \mu_2, s_1^2, R_1, R_2 \geq 0
\end{aligned}
\tag{8.29}
$$

The initial simplex table to the given problem is constructed and shown in Table 8.1.

In Table 8.1, all of the values of $(C_j - Z_j)$ are neither equal to nor less than zero. Therefore, optimality is not reached. The key column, ratio of *solution* and *value in the key column*, key row, and key element are evaluated and are shown in the same table. The maximum value for $(C_j - Z_j)$ is six corresponding to both columns x_1 and x_2. Either of these two variables can be entered into the basis. Considering that the entering variable is x_1 and the exiting variable is R_1, the corresponding key operations are shown in Table 8.2. The key column and key rows are shaded in gray.

In Table 8.2, all of the values of $(C_j - Z_j)$ are neither equal to nor less than zero. Therefore, optimality is not reached. The key column, ratio of *solution* and *value in the key column*, key row, and key element are evaluated and are shown in the same table. Variable x_2 is entered in the basis replacing the variable s_1^2. The corresponding key operations are shown in Table 8.3.

TABLE 8.1

Initial Simplex Table (Example 8.1)

CB_i	C_j	0	0	0	0	0	0	-1	-1		
	Basic Variable (BV)	x_1	x_2	λ_1	μ_1	μ_2	s_1^2	R_1	R_2	Solution	Ratio
-1	R_1	4	2	1	-1	0	0	1	0	4	1[a]
-1	R_2	2	4	2	0	-1	0	0	1	6	3
0	s_1^2	1	2	0	0	0	1	0	0	2	2
	Z_j	-6	-6	-3	1	1	0	-1	-1	-10	
	$C_j - Z_j$	6[b]	6	3	-1	-1	0	1	1		

[a] Key row; 4 is the key element.
[b] Key column.

In Table 8.3, all of the values of $(C_j - Z_j)$ are neither equal to nor less than zero. Therefore, optimality is not reached. Variable λ_1 is entered in the basis replacing the variable R_2. The corresponding key operations are shown in Table 8.4.

TABLE 8.2

Iteration Table 1 (Example 8.1)

	C_j	0	0	0	0	0	0	−1	−1		
CB_i	Basic Variable (BV)	x_1	x_2	λ_1	μ_1	μ_2	s_1^2	R_1	R_2	Solution	Ratio
0	x_1	1	1/2	1/4	−1/4	0	0	1/4	0	1	2
−1	R_2	0	3	3/2	1/2	−1	0	−1/2	1	4	4/3
0	s_1^2	0	3/2	−1/4	1/4	0	1	−1/4	0	1	2/3[a]
	Z_j	0	−3	−3/2	−1/2	1	0	1/2	−1	−4	
	$C_j - Z_j$	0	3[b]	3/2	1/2	−1	0	−3/2	0		

[a] Key row; 3/2 is the key element.
[b] Key column.

TABLE 8.3

Iteration Table 2 (Example 8.1)

	C_j	0	0	0	0	0	0	−1	−1		
CB_i	Basic Variable (BV)	x_1	x_2	λ_1	μ_1	μ_2	s_1^2	R_1	R_2	Solution	Ratio
0	x_1	1	0	1/3	−1/3	0	−1/3	1/3	0	2/3	2
−1	R_2	0	0	2	0	−1	−2	−1	1	2	1[a]
0	x_2	0	1	−1/6	1/6	0	2/3	−1/6	0	2/3	−
	Z_j	0	0	−2	0	1	2	1	−1	−2	
	$C_j - Z_j$	0	0	2[b]	0	−1	−2	−2	0		

[a] Key row; 2 is the key element.
[b] Key column.

TABLE 8.4

Iteration Table 3 (Example 8.1)

	C_j	0	0	0	0	0	0	−1	−1	
CB_i	Basic Variable (BV)	x_1	x_2	λ_1	μ_1	μ_2	s_1^2	R_1	R_2	Solution
0	x_1	1	0	0	−1/3	1/6	0	1	−1/6	1/3
0	λ_1	0	0	1	0	−1/2	−1	−1/2	1/2	1
0	x_2	0	1	0	1/6	−1/12	1/2	−1/4	1/12	5/6
	Z_j	0	0	0	0	0	0	0	0	0
	$C_j - Z_j$	0	0	0	0	0	0	−1	−1	

In Table 8.4, all of the values of $(C_j - Z_j) \leq 0$; hence, optimality is reached, and the corresponding optimal solution is as follows:

$$x_1 = \frac{1}{3} = 0.33; x_2 = \frac{5}{6} = 0.83; \lambda_1 = 1; \mu_1 = 0; \mu_2 = 0; s_1^2 = 0$$

This solution satisfies the complementary conditions (i.e., $\lambda_1 s_1 = 0$; $\mu_1 r_1 = 0$; $\mu_2 r_2 = 0$) and the restrictions of the signs of Lagrangian multipliers (i.e., λ_1, μ_1, μ_2). Further, the objective function is zero, which implies that the current solution is also feasible.

Therefore, the maximum value of the given quadratic problem is

$$Z(\max) = 4x_1 + 6x_2 - 2x_1^2 - 2x_1x_2 - 2x_2^2$$
$$= 4 \times \left(\frac{1}{3}\right) + 6 \times \frac{5}{6} - 2 \times \left(\frac{1}{3}\right)^2 - 2 \times \frac{1}{3} \times \frac{5}{6} - 2 \times \left(\frac{5}{6}\right)^2 = \frac{25}{6} = 4.17$$

The graphical representation of the optimization is shown in Figure 8.1. The point with the coordinates (0.33, 0.83) is the optimal point with maximum Z value of 4.17.

The MATLAB® coding for the aforesaid optimization problem using quadratic programming is given in Section 8.4.

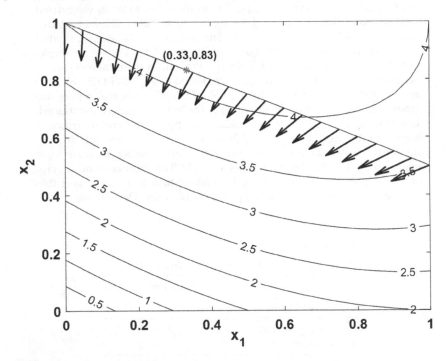

FIGURE 8.1
Graphical representation of the optimization problem (Example 8.1).

8.3 Application of Wolfe's Method for Solving Quadratic Programming Problem in Textile Field

An example of an application of Wolfe's method (Wolfe 1959) for solving a QPP in the textile field is demonstrated in Example 8.1.

EXAMPLE 8.2

In this example, the experimental data of Majumdar et al. (2016) are considered for optimization of thermal conductivity with desired air permeability and ultraviolet protection factor (UPF) of 1×1 rib-knitted fabrics using QPP. Majumdar et al. (2016) have taken 100% cotton yarns of three different fineness (5, 7.5, and 10 Ne) to produce 1×1 rib-knitted fabrics. All of the samples were prepared in a 12-gauge computerized flat knitting machine. In this QPP example, only two treatments (loop length and yarn count) each of three levels are considered. A full factorial experimental design plan for two factors and three levels gives a total of nine experimental runs (i.e., $3^2 = 9$) and four replications resulting in $9 \times 4 = 36$ samples. Table 8.5 shows the actual values of the factors corresponding to their coded levels where the controlled factors x_1 and x_2 correspond to loop length and yarn count, respectively. All of the 36 rib fabrics were washed in a Wascator washing machine as per the EN ISO 6330:2012 standard for complete relaxation. The samples were dried and then conditioned at a standard temperature of $20 \pm 2°C$ and relative humidity of $65 \pm 4\%$ for 48 hours. The samples were then evaluated for air permeability, thermal conductivity, and UPF. For each sample, 10 tests were performed, and their average values were taken. The air permeability of the knitted samples was tested using TEXTTEST FX 3000 air permeability tester as per the ASTM D737:2012 standard. A pressure gradient of 100 Pa was maintained as per the testing standard mentioned. Thermal conductivity was measured using Alambeta instrument according to the ISO EN 31092:2014 standard. As per this testing standard, the measuring head temperature and contact pressure were maintained at 32°C and 200 Pa, respectively. The thermal conductivity of a textile material is its ability to conduct heat through it, and this depends on the material and the porosity (Dias and Delkumburewatte,

TABLE 8.5

Actual and Coded Levels of Controlled Factors

Controlled Factors	Coded Level		
	−1	0	+1
Loop length (x_1), mm	5.09	5.39	5.69
Yarn count (x_2), Ne	5	7.5	10

Source: Majumdar, A. et al. 2016. *Journal of the Textile Institute*, 108(1), 110–116.

2008; Oglakcioglu et al., 2009; Chidambaram et al., 2011; Mal et al., 2016). The thermal conductivity of a material can be expressed as

$$\text{Thermal conductivity}(\lambda) = \frac{Qh}{A\Delta Tt}\,\text{W/mK} \tag{8.30}$$

where Q is quantity of heat conducted (J), h is thickness of the material (m), A is the area of the surface (m^2), Δt is the temperature differential across the material (K), and t is the time (s).

The UPF tests were determined by *in vitro* method as per AATCC 183: 2004 standard using Labsphere 2000F. The UPF is a rating that indicates how effectively a fabric blocks ultraviolet rays (AS/NZS 4399, 1996; ASTM D6603-07, 2007). The UPF of a fabric sample was calculated using the following equation:

$$\text{UPF} = \frac{\sum_{290}^{400} E(\lambda)S(\lambda)\Delta(\lambda)}{\sum_{290}^{400} E(\lambda)S(\lambda)T(\lambda)\Delta(\lambda)} \tag{8.31}$$

where $E(\lambda)$ is the relative erythemal spectral effectiveness, $S(\lambda)$ is the solar spectral irradiance [W/m²nm¹], $\Delta\lambda$ is the measured wavelength interval [nm], and $T(\lambda)$ is the average spectral transmittance of the sample.

The experimental results of 36 knitted samples are given in Table 8.6 (Majumdar et al., 2016).

The following equations were obtained for the quality parameters, where loop length (x_1) and yarn count (x_2) are in coded level:

$$\text{Thermal conductivity} = 52.38 - 3.7x_1 - 10.18x_2$$
$$+ 0.79x_1x_2 - 0.31x_1^2 + 2.23x_2^2 \tag{8.32}$$

$$\text{Air permeability} = 58.05 + 13.05x_1 + 42.33x_2 \tag{8.33}$$

$$\text{UPF} = 75.78 - 19.45x_1 - 53.97x_2 \tag{8.34}$$

These mathematical equations for thermal conductivity, air permeability, and UPF are used to design a knitted fabric for a cold climate. A fabric for a cold climate requires low thermal conductivity to resist dissipation of heat from the body through conduction. Also, low air permeability of a fabric means less transmission of cold air from the environment through the pores of the fabric to the body. So, for fabrics for cold climatic conditions, it is desirable to minimize either air permeability or thermal conductivity or both for better comfort. In this example, the thermal conductivity for rib fabric has been considered as the objective function to be maximized, and the air permeability and UPF were taken as constraint functions to be retained at a desired level. Accordingly, the optimization problem for 1×1 rib fabric has been formulated as follows:

TABLE 8.6

Experimental Results of Knitted (1 × 1 rib) Fabrics

Experiment Number	Loop Length in mm (x_1)	Yarn Count in Ne (x_2)	Thermal Conductivity (W/mK × 10^{-3})	Air Permeability (cm³/cm²/s)	UPF
1	5.09	7.5	57.6	36.7	75.1
2	5.09	7.5	56.8	38.7	74.3
3	5.09	7.5	57.1	38.1	76.0
4	5.09	7.5	56.7	35.1	79.5
5	5.69	7.5	46.3	61.6	52.2
6	5.69	7.5	47.9	63.8	48.1
7	5.69	7.5	46.3	67.7	45.4
8	5.69	7.5	46.1	61.1	59.2
9	5.39	7.5	52.7	53.2	60.0
10	5.09	5	67.0	10.3	206.0
11	5.09	10	46.1	85.2	34.1
12	5.09	5	69.1	14.6	170.7
13	5.09	10	45.4	88.5	30.3
14	5.69	5	62.8	29.9	88.9
15	5.69	10	40.8	124.9	22.9
16	5.69	5	60.6	26.9	97.7
17	5.69	10	42.2	116.3	23.7
18	5.39	7.5	52.5	43.5	63.7
19	5.09	5	68.4	15.1	165.2
20	5.09	10	46.0	83.3	35.5
21	5.09	5	70.1	14.5	185.6
22	5.09	10	46.7	82.3	34.8
23	5.69	5	59.1	25.0	116.5
24	5.69	10	42.1	128.8	22.2
25	5.69	5	62.2	26.4	101.7
26	5.69	10	41.9	123.1	21.9
27	5.39	7.5	52.1	49.2	58.6
28	5.39	5	64.7	20.7	120.7
29	5.39	10	43.8	105.2	28.6
30	5.39	5	65.2	19.4	126.0
31	5.39	10	43.5	105.6	28.3
32	5.39	5	63.3	19.5	124.7
33	5.39	10	46.9	107.3	29.4
34	5.39	5	62.5	15.5	133.2
35	5.39	10	45.2	103.3	29.9
36	5.39	7.5	54.0	49.9	57.5

Source: Majumdar, A. et al. 2016. *Journal of the Textile Institute,* 108(1), 110–116.

$$\text{Minimize:} \quad \text{Thermal conductivity}$$
$$\text{Subject to:} \quad \text{Air permeability} \leq 50$$
$$\text{UPF} \geq 25$$

Thus, the optimization problem becomes

Minimize $Z = 52.38 - 3.7x_1 - 10.18x_2 + 0.79x_1x_2 - 0.31x_1^2 + 2.23x_2^2$
subject to: $58.05 + 13.05x_1 + 42.33x_2 \leq 50$
$$75.78 - 19.45x_1 - 53.97x_2 \geq 25 \tag{8.35}$$

After rearranging, the constraint functions become

$$13.05x_1 + 42.33x_2 \leq -8.05$$

$$19.45x_1 + 53.97x_2 \leq 50.78$$

The solution of this optimization problem is as follows:

$$x_1 = -0.6592, \text{ which corresponds to 5.19 mm}$$

$$x_2 = 0.013, \text{ which corresponds to 7.53 Ne}$$

The minimum value of the objective function is 54.545.

The graphical representation of the optimization is shown in Figure 8.2. The point with the coordinates (-0.6592, 0.013) is the optimum point that corresponds to $x_1 = 5.19$ mm and $x_2 = 7.53$ Ne with a minimum Z (thermal conductivity) value of 54.545 W/mK $\times$ 10^{-3}.

The MATLAB® coding for the aforesaid optimization problem using the quadratic programming method is given in Section 8.4.

8.4 MATLAB® Coding

MATLAB® Coding for the Optimization Problem Given in Example 8.1 Using Quadratic Programming

```
clc
clear all
close all
 H=[4 2;2 4];
f=[-4 -6]';
A=[1 2];
b=[2];
lb = [0 0]';
X0=[0.5 0.5];
```

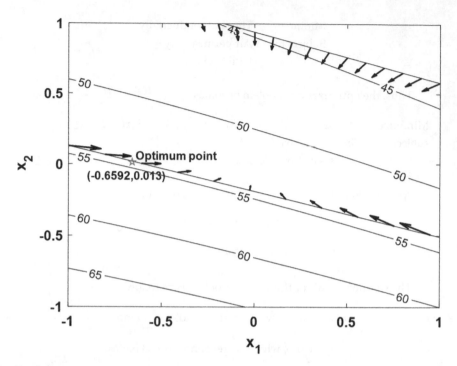

FIGURE 8.2
Graphical representation of the optimization problem (Example 8.2).

```
options = optimoptions('quadprog','Algorithm','interior-
point-convex','Display','off');
[x,fval,output,lambda] = quadprog(H,f,A,b,[],[],lb,[],X0,
options)
max_val=-fval
xx1=0:0.05:1;
yy1=0:0.05:1;
[X,Y]=meshgrid(xx1,yy1);
Z=(4*X+6*Y-2*X.^2-2*X.*Y-2*Y.^2);
[C,h]=contour(X,Y,Z,'color','k','LineWidth',0.5);
clabel(C,h,'FontSmoothing','on');
hold on
x2=0:0.05:1;
y2=(2-x2)/2;
plot(x2,y2,'k')
hold on
plot(x(1),x(2),'*r')
hold on
quiver(x2,y2,-x2,-y2,0.5,'k','LineWidth',1.3)
v1=x(1)+2*x(2)
z_val= 4*x(1)+6*x(2)-2*x(1)^2-2*x(1)*x(2)-2*x(2)^2
xlim([0 1])
```

```
ylim([0 1])
set(gcf,'color', 'W')
xlabel('x_{1}')
ylabel('x_{2}')
```

MATLAB® Coding for the Optimization Problem Given in Example 8.2 Using Quadratic Programming

```
clc
clear all
close all
H=[-0.62 0.79;0.79 4.46];
f=[-3.7 -10.18]';
A=[13.05 42.33;19.45 53.97];
b=[-8.05;50.78];
lb = [-1 -1]';
ub = [1 1]';
X0=[0.5 0.5];
[x,fval,output,lambda] = quadprog (H,f,A,b,[],[],lb,ub,X0,[])
x1=-1:0.1:1;
y1=-1:0.1:1;
[X1,Y1] = meshgrid(x1,y1);
Z=52.38 - 3.7*X1 - 10.18*Y1+0.79*X1.*Y1-0.31*X1.^2+2.23*Y1.^2;
[C,h] = contour(X1,Y1,Z,'color','k','LineWidth',0.5);
clabel(C,h,'FontSmoothing','on');
hold on
x2=-1:0.1:1;
y2=(-8.05-13.45*x2)/42.33;
plot(x2,y2,'k')
hold on
x3=-0.5:0.1:1;
y3=(50.78-19.45*x3)/53.97;
plot(x3,y3,'k-')
hold on
scatter(x(1),x(2),'pr')
hold on
x2=-1:0.195:1;
y2=(-8.05-13.45*x2)/42.33;
quiver(x2,y2,-x2,-y2,0.3,'k','LineWidth',1.3)
hold on
quiver(x3,y3,-x3,-y3,0.3,'k','LineWidth',1.3)
v1=13.05*x(1)+42.33*x(2)
v2=19.45*x(1)+53.97*x(2)
Z_val=52.38- 3.7*x(1) - 10.18*x(2)+0.79*x(1)*x(2)-0.31*x(1)^2
+2.23*x(2)^2
xlim([-1 1])
ylim([-1 1])
 set(gcf,'color', 'W')
xlabel('x_{1}')
ylabel('x_{2}')
```

8.5 Summary

LP process models are used when the optimization problem has linear equations. When the optimization problem is complex, a nonlinear programming method is used to solve it. In a QPP, the objective function is quadratic, and constraints are linear. Solving a QPP requires conversion of the quadratic objective function and linear constraints into a Lagrangian function using Lagrangian multipliers. In the beginning of this chapter, the formation of the Lagrangian function is explained. Then the Kuhn-Tucker conditions for optimality are discussed. The Kuhn-Tucker conditions for constrained optimization problems are calculated based on unconstrained optimization of the corresponding Lagrangian function. The optimal point for the optimization problems is the point that satisfies Kuhn-Tucker conditions. Wolfe's method has been demonstrated to solve QPPs. A simple example of quadratic programming problem-solving using Wolfe's method is demonstrated step by step. Last, an example of the application of quadratic programming in the textile field is presented.

References

AATCC 183. 2004. *Transmittance or Blocking of Erythemally Weighted Ultraviolet Radiation through Fabrics*, American Association of Textile Chemists and Colorists, Research Triangle Park, NC.

AS/NZS 4399. 1996. *Sun Protective Clothing—Evaluation and Classification*, Australian/New Zealand Standards, Sydney, NSW.

ASTM D6603. 2007. *Standard Guide for Labelling of UV-Protective Textiles*, ASTM International, West Conshohocken, PA.

ASTM D737. 2012. *Standard Test Method for Air Permeability of Textile Fabrics*, ASTM International, West Conshohocken, PA.

Chidambaram, P., Govind, R. and Venkataraman, K. C. 2011. The effect of loop length and yarn linear density on the thermal properties of bamboo knitted fabrics. *AUTEX Research Journal*, 11(4), 102–105.

Deb, K. 2010. *Optimization for Engineering Design*, PHI Learning, New Delhi, India.

Dias, T. and Delkumburewatte, G. B. 2008. Changing porosity of knitted structures by changing tightness. *Fibres and Polymers*, 9(1), 76–79.

EN ISO 6330. 2012. *Domestic Wash and Dry Procedures for Textiles*, Comite Europeen de Normalisation.

EN ISO 31092. 2014. *Textiles – Physiological Effects – Measurement of Thermal and Water-Vapor Resistance under Steady-State Conditions (Sweating Guarded-Hotplate Test)*, Comite Europeen de Normalisation.

Kuhn, H. W. and Tucker, A. W. 1951. Nonlinear programming. *Proceedings of Second Berkeley Symposium on Mathematical Statistics and Probability*. University of California Press, Berkeley, CA. 481–492.

Majumdar, A., Mal, P., Ghosh, A. and Banerjee, D. 2016. Multi-objective optimation of air permeability and thermal conductivity of knitted fabrics with desired ultraviolet protection. *Journal of the Textile Institute*, 108(1), 110–116.

Mal, P., Ghosh, A., Majumdar, A. and Banerjee, A. 2016. Engineering of knitted cotton fabrics for optimum comfort in a hot climate. *Fibres and Textiles in Eastern Europe*, 24, 102–106.

Oglakcioglu, N., Celik, P., Bedez Ute, T., Marmarali, A. and Kadoglu, H. 2009. Thermal comfort properties angora rabbit/cotton fiber blended knitted fabrics. *Textile Research Journal*, 79(10), 888–894.

Panneerselvam, R. 2002. *Operations Research*, Prentice Hall, New Delhi, India.

Sharma, J. K. 2010. *Operations Research, Theory and Applications*, Macmillan, New Delhi, India.

Wolfe, P. 1959. The simplex method for quadratic programming. *The Economic Society, Econometrica*, 27(3), 382–398.

9

Genetic Algorithm

9.1 Introduction

The term *optimization* may be defined as the process of obtaining the best possible result out of all the conditions and available means. There is scarcely any area of scientific concern where optimization has not made its dent. With the advent of high-speed computational techniques, optimization algorithms have empowered scientists and technologists from diverse engineering disciplines where the emphasis is on maximizing or minimizing a certain objective. Focusing attention in the sphere of textile engineering, a rapidly increasing use of the optimization technique is now well recognized in machine design, process parameter setting, and engineering design of fibers, yarns, fabrics, and composites.

The field of optimization has changed over the last few decades with the introduction of a number of nontraditional optimization algorithms that are based on the concept of natural phenomena. Of these, the genetic algorithm mimics nature's evolutionary principles to drive its search toward an optimal solution. One of the most striking differences between the genetic algorithm and classical optimization algorithms is that the latter use a point-by-point approach, where one solution in each iteration is modified to a different (hopefully better) solution, and eventually the outcome becomes a local optimized solution, whereas the genetic algorithm works with a population of solutions in each iteration, instead of a single solution. As multiple solutions are processed concomitantly, it is likely that the expected final solution of the genetic algorithm may be a global optimized solution.

The genetic algorithm is a heuristic search algorithm that proceeds by randomly generating an initial population of individuals. Each individual is represented by a binary-coded string or chromosome encoding a possible solution in the data space. At every generation or iteration step, the individuals in the current population are tested according to the fitness function. To form a new population (the next generation), individuals are selected according to their fitness. Selection alone cannot introduce new individuals into the population, which is necessary in order to make the solution as independent of the initial population as possible. New individuals in the search space are thus generated by two operations: crossover and mutation. Crossover concerns two selected individuals (parents) who exchange parts of their chromosome to form two new individuals (offsprings). Mutation consists of flipping bits of an individual's strings at random and with some small probability. In fact, new individuals in the search space are sampled randomly, thereby leading to a global optimized solution rather than a local optimized solution.

9.2 Genetic Algorithm

The genetic algorithm is a population-based probabilistic optimization technique. It mimics Darwin's principle of natural selection based on the "survival of the fittest." John Holland (1975) of the University of Michigan, Ann Arbor, was the first to propose the concept of the genetic algorithm. Thereafter, a number of scientists, namely, Goldberg (1989), Davis (1991), and Deb (2005), have contributed to developing this subject. The main operations of the genetic algorithm are explained in the rest of this chapter.

9.2.1 Representation

The genetic algorithm starts with a population of randomly generated initial solutions of size N, which are represented by binary strings. A binary string is analogous to a biological chromosome, and each bit of the string is nothing but a gene value. The optimization problem is defined in the real world, and the encoding or representation of the variables is done in the binary world as displayed in Figure 9.1. The operators, like crossover and mutation, are applied

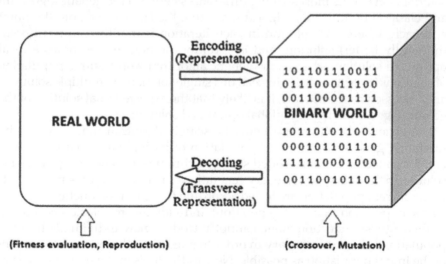

FIGURE 9.1
Representation.

in the binary world. By means of decoding or transverse representation, we go back to the real world where fitness evaluation and reproduction operations are performed.

9.2.2 Fitness Evaluation

Because the genetic algorithm is based on the concept of "survival of the fittest," it can solve only a maximization problem. Hence, minimization of a function $f(x)$ has to be converted into a suitable maximization problem either by maximizing $-f(x)$ or maximizing $1/f(x)$ for $f(x) \neq 0$, or maximizing $1/1 + f(x)$ for $f(x) \neq -1$, etc. The fitness values or the values of the objective function of a maximization problem are evaluated for each string in the population. In order to determine the fitness value of a binary string, the real values of the variables are to be estimated first. A binary coded string can be converted into real values of variables by means of linear mapping using the following expression:

$$x_i = x_{i(\min)} + \frac{x_{i(\max)} - x_{i(\min)}}{2^{l_i} - 1} \times D_i \tag{9.1}$$

where x_i is the real value; $x_{i(\max)}$ and $x_{i(\min)}$ are the maximum and minimum values; l_i is the string length; D_i is the decoded value corresponding to the ith variable; and $i = 1, 2, \ldots, m$, where m is the number of variables.

9.2.3 Reproduction

All of the strings in a population may not be equally good with respect to their fitness values. The reproduction operator is utilized to choose the good strings using their fitness values. Thus, it forms a mating pool consisting of good solutions probabilistically. It is important to note that the mating pool may contain multiple copies of a particular good solution. The size of the mating pool is kept equal to that of the population of solutions considered before reproduction. Thus, the average fitness of the mating pool is expected to be higher than the average fitness of the population before reproduction. There exists a number of reproduction schemes, namely, Roulette-wheel selection, tournament selection, ranking selection, etc. which are briefly explained in the following sections.

Roulette-Wheel Selection

In roulette-wheel selection, the probability of a string being selected in the mating pool depends on its fitness value. A string with a higher fitness value has a greater probability of being selected in the mating pool. The working principle of roulette-wheel selection is explained with the help of Figure 9.2. Let us suppose that there are only three strings A, B, and C in the population with their fitness values 1, 2, and 3, respectively. A summation of the fitness values turns out to be 6. Thus, the selection probabilities of

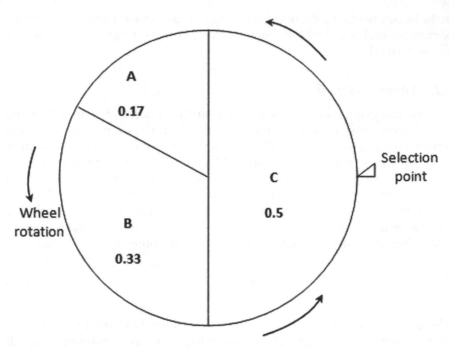

FIGURE 9.2
Roulette-wheel selection.

strings A, B, and C in the mating pool are 0.17 (1/6), 0.33 (2/6), and 0.5(3/6), respectively, which gives the cumulative probability equal to 1. Now we design a roulette wheel that has total area of 1 and assign the three strings to the wheel in such a way that area shared by each string is proportional to its selection probability. Thus, it is evident from Figure 9.2 that string C shares the highest area (50%), and string A shares the lowest area (17%) in the roulette wheel. The wheel is rotated three times, and each time a fixed selection point indicates the winning string, after it is stopped. Let us assume that string C wins twice and string B wins once. Hence, string C is copied twice, and string B is copied once in the mating pool, whereas because string A is weaker, it is eliminated.

Tournament Selection

In this method, a tournament is played by selecting a small number of strings of size n at random from the population of size N, and the best one is determined on the basis of fitness value. The size n is kept much smaller than N. After a tournament is played, the best string gets copied into the mating pool, and then all n strings are returned to the population. Hence, only one string is copied per tournament. Subsequently, N number of tournaments is to be played to ensure the size of the mating pool equals N.

Ranking Selection

In the ranking selection method, first the strings are arranged in ascending order of their fitness values. The strings having the worst fitness are assigned to rank 1, second worst fitness to rank 2, etc., and the best fitness will have rank N (number of population). In the next step, a proportionate selection scheme based on the assigned rank is adopted. The working principle of ranking selection is explained with the help of the following example. Let us suppose that there are only three strings X, Y, and Z in the population whose fitness values are 15, 4, and 1, respectively. Accordingly, the strings Z, Y, and X will be assigned to ranks 1, 2, and 3, respectively. If a fitness-based proportionate selection (roulette-wheel selection) would have been carried out, the probabilities of being selected in the mating pool for strings X, Y, and Z would become 0.75, 0.2, and 0.05, respectively, as depicted in Figure 9.3.

However, in case of ranking selection, the probability of being selected for a string in the mating pool is determined from the expression $r_i/\Sigma r_i$, where r_i indicates the rank of the ith string. Thus, the selection probabilities of the strings X, Y, and Z in the mating pool become 0.5, 0.33, and 0.17, respectively, as shown in Figure 9.4. Rank-based proportionate selection can avoid premature convergence, but it may lead to slower convergence, because the best strings do not differ so much from other ones.

9.2.4 Crossover

In crossover, an exchange of properties between the parents is taking place which leads to the creation of new children solutions. Thus, the purpose of

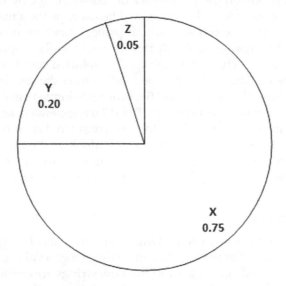

FIGURE 9.3
Fitness-based proportionate selection.

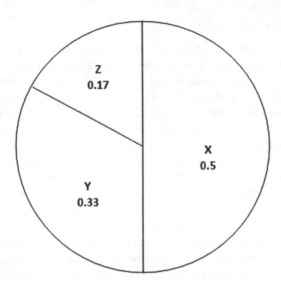

FIGURE 9.4
Rank-based proportionate selection.

the crossover operation is to explore the search space by discovering the promising area.

To begin with the crossover, first the mating pairs (each pair consists of two parent strings) are formed from the mating pool. If the population size is N, then $N/2$ mating pairs are formed. The selection of a mating pair to participate in crossover depends on the probability of crossover (p_c). By tossing a coin, the probability of success or "heads" is p_c; if heads appears, a mating pair will participate in crossover to produce two offspring, otherwise, no crossover will take place and the mating pool will remain intact in the population. If p_c is too large, then the structure of a high-quality solution could be prematurely destroyed; on the contrary, a p_c that is too small reduces the searching efficiency. Generally, p_c is chosen between 0.6 and 0.9. In order to implement the coin tossing artificially, a random number between 0.0 and 1.0 is generated, and if it is found to be smaller than or equal to p_c, the outcome is considered as success or "heads."

Various types of crossover operators are available in the literature, such as single-point crossover, two-point crossover, multipoint crossover, uniform crossover, etc., which are briefly discussed in the following sections.

Single-Point Crossover

In this scheme, a crossover site (j) lying between 1 and ($l - 1$) is randomly selected, where l specifies the length of the string. Once a value of j is obtained, the strings are crossed at the site j and two new strings are created. In general, crossing is done between the two substrings lying on the right side of the crossover site. An example of single-point crossover is shown as follows, where the first parent string is represented by all zeros and the second parent string is represented by all ones:

$$0\ 0\ 0\ 0\ 0\ 0\ 0\ 0\ 0\ 0|0\ 0\ 0\ 0\ 0\ 0$$

$$1\ 1\ 1\ 1\ 1\ 1\ 1\ 1\ 1\ 1|1\ 1\ 1\ 1\ 1\ 1$$

Here the length of the string is 16. Thus, a number between 1 and 15 is created at random to choose the crossing site. Suppose that the obtained random number is 10. That means that we cross the strings at site 10 to produce two new strings as follows:

$$0\ 0\ 0\ 0\ 0\ 0\ 0\ 0\ 0\ 0|1\ 1\ 1\ 1\ 1\ 1$$

$$1\ 1\ 1\ 1\ 1\ 1\ 1\ 1\ 1\ 1|0\ 0\ 0\ 0\ 0\ 0$$

Two-Point Crossover

In this scheme, two different crossover sites lying between 1 and $(l-1)$ are selected at random, and the bits inside the two crossover sites are swapped between the parent strings. The parent strings participating in crossover are shown as follows, where random points 4 and 9 are chosen as two crossing sites:

$$0\ 0\ 0\ 0|0\ 0\ 0\ 0\ 0\ |0\ 0\ 0\ 0\ 0\ 0\ 0$$

$$1\ 1\ 1\ 1|1\ 1\ 1\ 1\ 1\ |1\ 1\ 1\ 1\ 1\ 1\ 1$$

The crossover is performed by swapping the bits inside 4 and 9 between the parent strings. Two children strings are thus produced due to the two-point crossover:

$$0\ 0\ 0\ 0|\ 1\ 1\ 1\ 1\ 1|0\ 0\ 0\ 0\ 0\ 0\ 0$$

$$1\ 1\ 1\ 1|0\ 0\ 0\ 0\ 0\ |1\ 1\ 1\ 1\ 1\ 1\ 1$$

Multipoint Crossover

In this scheme, multiple (more than two) crossover sites are selected at random along the length of the string. The bits lying inside the alternate pairs of crossover sites are swapped between the parent strings. The parent strings are shown as follows, where random points 3, 8, and 14 are chosen as three crossover sites:

$$0\ 0\ 0|0\ 0\ 0\ 0\ 0\ |0\ 0\ 0\ 0\ 0\ 0\ |0\ 0$$

$$1\ 1\ 1|1\ 1\ 1\ 1\ 1\ |1\ 1\ 1\ 1\ 1\ 1|1\ 1$$

The bits lying inside crossover sites 3 and 8 and the bits lying to the right side of crossover site 14 are swapped between the parent strings, while the remaining bits are kept unaltered. The generated children strings thus produced due to the multipoint crossover are as follows:

$$0\,0\,0|\;1\,1\,1\,1\,1\;|\;0\,0\,0\,0\,0\,0\;|\;1\,1$$

$$1\,1\,1|\;0\,0\,0\,0\,0|\;1\,1\,1\,1\,1\,1|\;0\,0$$

If the five crossover sites selected at random are 2, 5, 7, 11, and 13, then the parent strings will look as follows:

$$0\,0|\;0\,0\,0|\;0\,0|\;0\,0\,0\,0|\;0\,0|\;0\,0\,0$$

$$1\,1|\;1\,1\,1|\;1\,1|\;1\,1\,1\,1|\;1\,1|\;1\,1\,1$$

In this case, the bits lying inside crossover sites 2 and 5, 7 and 11, and the right side of 13 are swapped between the parent strings by keeping the other bits unaltered. Thus, the children strings produced after crossover will look as follows:

$$0\,0|\;1\,1\,1|\;0\,0|\;1\,1\,1\,1|\;0\,0|\;1\,1\,1$$

$$1\,1|\;0\,0\,0|\;1\,1|\;0\,0\,0\,0|\;1\,1|\;0\,0\,0$$

Uniform Crossover

In this scheme, at each bit position of the parent strings, a coin is tossed with a probability of appearing heads equal to 0.5 to determine whether there will be swapping of the bits. If head (H) appears at a bit position, there will be swapping of the bits between the parent strings. But if the outcome is "tails" (T), they will remain unaltered. In order to explain the principle of uniform crossover, let us take the same parent strings as in previous cases as follows:

$$0\,0\,0\,0\,0\,0\,0\,0\,0\,0\,0\,0\,0\,0\,0\,0$$

$$1\,1\,1\,1\,1\,1\,1\,1\,1\,1\,1\,1\,1\,1\,1\,1$$

Suppose that heads appear at 3rd, 6th, 7th, 10th, 11th, and 16th bit positions. After swapping the bits between parent strings at these selected positions, the children strings produced after crossover look as follows:

$$0\,0\,1\,0\,0\,1\,1\,0\,0\,1\,1\,0\,0\,0\,0\,1$$

$$1\,1\,0\,1\,1\,0\,0\,1\,1\,0\,0\,1\,1\,1\,1\,0$$

9.2.5 Mutation

In biology, mutation refers to a certain change in parameters on the gene level. In a genetic algorithm, the purpose of the mutation operation is to create a solution in the vicinity of the current solution; thereby, it brings a local change over the current solution. Thus, if a solution gets stuck at the local minimum, mutation may help it to come out of this situation; consequently, it may jump to the global basin. Hence, mutation helps to exploit the search space by optimizing within a promising area.

In mutation, 1 is converted into 0 and vice versa. To implement a bit-wise mutation scheme, we toss a coin with a probability of mutation (p_m) for every bit of a crossed-over pool. The value of p_m is generally kept low, because with a higher value of p_m, the genetic evolution degenerates into a random local search. If the outcome of the toss in a particular bit position is true, then that bit will be mutated; that is, 1 will be converted into 0 and vice versa. For this purpose, a random number lying between 0.0 and 1.0 is generated at every bit position, and if it is found to be smaller or equal to p_m, the outcome is considered as true. In order to explain the principle of mutation, let us consider that before mutation two strings look as follows:

$$0 \; 0 \; \boxed{0} \; 0 \; 0 \; 0 \; 0 \; 0 \; 0 \; 0 \; 0 \; \boxed{0} \; 0 \; 0 \; 0 \; 0$$

$$1 \; 1 \; 1 \; 1 \; 1 \; 1 \; \boxed{1} \; 1 \; 1 \; 1 \; 1 \; 1 \; 1 \; 1 \; 1 \; 1$$

While performing the toss in every bit position, suppose that the outcomes are true in 3rd and 12th positions for the first string and 7th position for the second string (marked by boxes). After mutation, the two strings look as follows:

$$0 \; 0 \; 1 \; 0 \; 0 \; 0 \; 0 \; 0 \; 0 \; 0 \; 0 \; 1 \; 0 \; 0 \; 0 \; 0$$

$$1 \; 1 \; 1 \; 1 \; 1 \; 1 \; 0 \; 1 \; 1 \; 1 \; 1 \; 1 \; 1 \; 1 \; 1 \; 1$$

Because of crossover, children acquire some common properties of their parents, but mutation introduces some new property for which children are a little bit different than their parents. Crossover leads to a big jump to an area somewhere in between two parent solutions, whereas mutation causes random small diversions near a solution. After some repeated crossover without mutation, the same solutions may repeat again and again. Thus, crossover alone may not be sufficient to reach the optimum solution. The role of mutation is explained with the help of Figure 9.5, which illustrates the plot of a bimodal function having one local minimum in the local basin and one global minimum in the global basin. Let us assume that all randomly

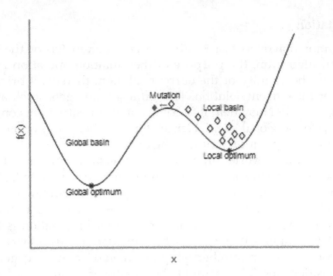

FIGURE 9.5
Bimodal function with one local minimum and one global minimum.

generated initial solutions fall on the local basin coincidentally. For example, consider the following population of the initial solution:

1 0 1 0 1 1 0 1

1 1 1 1 0 0 1 0

1 0 1 0 1 1 1 0

$\vdots$

1 1 0 1 0 0 0 1

It is noticed that all strings of the initial population have a 1 in their leftmost bit position. If the global optimum solution requires 0 in the leftmost bit position, then the crossover operator alone cannot create 0 in that position. In this situation, the mutation operator can only turn 1 into 0, thereby creating the possibility of pushing a string from the local basin into the global basin as depicted in Figure 9.5. Hence, mutation helps to achieve the global optimum solution by maintaining small diversity in the population.

9.2.6 Flowchart of a Genetic Algorithm

The flowchart of a genetic algorithm is shown in Figure 9.6. There are six steps as follows:

- *Step 1*: Initialize population size, maximum number of generations (gen_{max}), crossover probability, mutation probability, lower and upper boundaries of variables, and bit size for each variable. Generate a random population of binary strings. Set gen $= 0$.

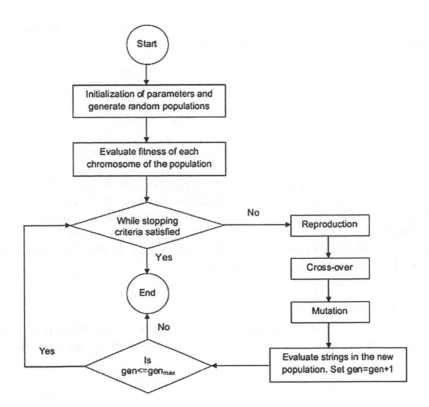

FIGURE 9.6
Flowchart of genetic algorithm program.

- *Step 2*: Evaluate the fitness of each string in the population.
- *Step 3*: If gen > gen$_{max}$ or other stopping criteria are satisfied, stop. Otherwise, carry out the reproduction operation.
- *Step 4*: Execute the crossover operation.
- *Step 5*: Perform the mutation operation.
- *Step 6*: Evaluate the strings in the new population, set gen = gen + 1, and go to step 3.

9.3 Step-by-Step Working Principle of Genetic Algorithm

Let us consider the following optimization problem to explain the working principle of the genetic algorithm:

$$\text{Minimize} \quad f(x_1, x_2) = 4x_1^2 + 3x_2^2 - 6x_1x_2 - 4x_1 + 24$$
$$\text{Subject to} \quad 0 \le x_1, x_2 \le 10 \tag{9.2}$$

where x_1 and x_2 are the real variables. As the genetic algorithm is designed for maximization, minimization of the previous function has been modified as

$$\text{Maximize} \quad F(x_1, x_2) = \frac{1}{1 + f(x_1, x_2)}$$

$$\text{Subject to} \quad 0 \le x_1, x_2 \le 10$$

(9.3)

Here we denote $f(x_1, x_2)$ and $F(x_1, x_2)$ as the objective function and fitness function, respectively. Surface and contour plots of the objective function $f(x_1, x_2)$ are displayed in Figures 9.7 and 9.8, respectively. It can be shown that the minimum value of the function $f(x_1, x_2)$ is 20, which is obtained at $x_1 = 2$ and $x_2 = 2$. In Figures 9.7 and 9.8, the optimum solution is marked by an asterisk (*).

The following steps are involved when solving the optimization problem of Equation 9.2 using the genetic algorithm.

Step 1: Generation of Initial Population

An initial population of solutions of size N is selected at random. The solutions are expressed in binary coding for representing the variables x_1

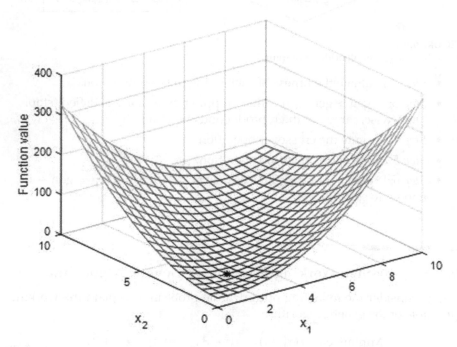

FIGURE 9.7
Surface plot of function $f(x_1, x_2) = 4x_1^2 + 3x_2^2 - 6x_1x_2 - 4x_1 + 24$.

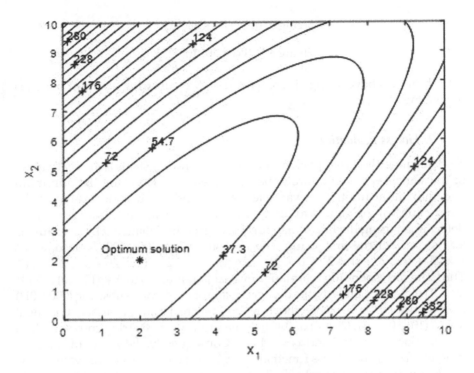

FIGURE 9.8
Contour plot of the function $f(x_1, x_2) = 4x_1^2 + 3x_2^2 - 6x_1x_2 - 4x_1 + 24$.

and x_2. We assigned 8 bits to represent each variable; hence, each string of the population is 16 bits long. The population size (N) may become 100, 200, ... , 1,000, depending on the complexity of the problem. However, for the sake of simplicity, let us suppose that the initial population of only 10 strings is created at random:

String 1: 1000110000001011

String 2: 0001010000100101

String 3: 0011111111111111

String 4: 0101111111110100

String 5: 1010011100000000

String 6: 0110110000111000

String 7: 0110000101101111

String 8: 0011001010010111

String 9: 0001111110110111

String 10: 0111001001001111

Each of the above strings is composed of two substrings, each of which is 8 bits long.

Step 2: Fitness Evaluation

To determine the fitness value of each solution (i.e., string), the real values of the variables x_1 and x_2 are to be calculated first. Knowing the minimum and maximum limits of a variable (say x_1) and the decoded value of the binary substring assigned to represent it, the real value of variable x_1 can be determined using the linear mapping rule given in Equation 9.1. For example, let us consider the first string. The first substring (10001100) is decoded as $1 \times 2^7 + 0 \times 2^6 + 0 \times 2^5 + 0 \times 2^4 + 1 \times 2^3 + 1 \times 2^2 + 0 \times 2^1 + 0 \times 2^0 = 140$. Thus, the corresponding real value of variable x_1 is equal to $0 + (10 - 0) \times 140/(2^8 - 1)$ or 5.4902. Similarly, the decoded value of the second substring (00001011) is worked out to be 11. Hence, the corresponding real value of variable x_2 is equal to $0 + (10 - 0) \times 11/(2^8 - 1)$ or 0.43137. Therefore, the first string corresponds to the solution $x_1 = 5.4902$ and $x_2 = 0.43137$. Consequently, the values of objective function $f(x_1,x_2)$ and fitness function $F(x_1,x_2)$ for the first string are evaluated as 108.96 and 0.009095, respectively. The same procedure is repeated to evaluate the fitness values of the other strings of the population. Table 9.1 illustrates the decoded values, real values of variables, objective function values, and fitness function values of all 10 strings in the initial population. From the last column of Table 9.1, the average and maximum values of fitness for the initial population can be estimated, which are 0.01888 and 0.041997, respectively. Therefore, the ratio (R) between the average fitness and maximum fitness of the

TABLE 9.1

Fitness Evaluation

String Number	Substring 1	Substring 2	D_1	D_2	x_1	x_2	$f(x_1,x_2)$	$F(x_1,x_2)$
1	10001100	00001011	140	11	5.4902	0.43137	108.96	0.009095
2	00010100	00100101	20	37	0.78431	1.451	22.811	0.041997
3	00111111	11111111	63	255	2.4706	10	190.3	0.005228
4	01011111	11110100	95	244	3.7255	9.5686	125.4	0.007911
5	10100111	00000000	167	0	6.549	0	169.36	0.00587
6	01101100	00111000	108	56	4.2353	2.1961	37.472	0.025993
7	01100001	01101111	97	111	3.8039	4.3529	24.158	0.039748
8	00110010	10010111	50	151	1.9608	5.9216	67.065	0.014692
9	00011111	10110111	31	183	1.2157	7.1765	127.21	0.0078
10	01110010	01001111	114	79	4.4706	3.098	31.755	0.030529

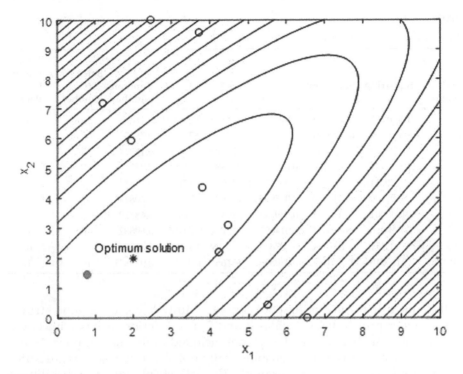

FIGURE 9.9
The initial population and the best individual on the contour plot of the objective function.

initial population becomes 0.4497. It appears from Table 9.1 that string number 2 gives the best fit among the initial population, and the corresponding values of x_1 and x_2 are 0.78431 and 1.451, respectively. In Figure 9.9, the initial populations are marked with the empty circles on the contour plot of the objective function, whereas the best individual (string 2) is marked with the filled circle.

Step 3: Reproduction

Here we used the roulette-wheel selection method for the purpose of parent selection. Table 9.2 depicts the working principle of roulette-wheel selection.

It begins with computation of the probability of selection (P) of each string on the basis of its fitness value. Thus, the value of (P) for each string is obtained as $F(x_1,x_2)/\Sigma F(x_1,x_2)$. Subsequently, the cumulative probability (Q) of each string is computed. In the next step, random numbers between 0 and 1 are created for spotting the particular string to be copied in the mating pool. For an example, the first random number generated as shown in Table 9.2 is 0.11005, which lies between two values of cumulative probability 0.048154 and 0.27052. As a result of that, the second string gets copied in the mating pool. In a similar manner, the other strings are copied in the mating pool. The last column of Table 9.2 shows the true count of each string in the mating pool.

TABLE 9.2

Roulette-Wheel Selection

String Number	Substring 1	Substring 2	$F(x_1,x_2)$	P	Q	Random Number	Selected String Number	True Count in Mating Pool
1	10001100	00001011	0.009095	0.048154	0.048154	0.11005	2	0
2	00010100	00100101	0.041997	0.22237	0.27052	0.78126	8	3
3	00111111	11111111	0.005228	0.027679	0.2982	0.15344	2	1
4	01011111	11110100	0.007911	0.041888	0.34009	0.2971	3	0
5	10100111	00000000	0.00587	0.03108	0.37117	0.3584	5	1
6	01101100	00111000	0.025993	0.13763	0.5088	0.20844	2	1
7	01100001	01101111	0.039748	0.21046	0.71926	0.43122	6	2
8	00110010	10010111	0.014692	0.077791	0.79705	0.69882	7	1
9	00011111	10110111	0.0078	0.041299	0.83835	0.67507	7	0
10	01110010	01001111	0.030529	0.16165	1	0.936871	10	1

It is noted that the mating pool contains multiple copies of good strings with higher probabilities such as 2 and 7. The inferior strings with lower probabilities such as 1, 4, and 9 are not included in the mating pool. As the selection of strings is made probabilistically, not all selected strings are better than all rejected strings. For example, the fifth string with a probability of 0.03108 is selected, but the first string with a probability of 0.048154 has not been selected. The strings in the mating pool after reproduction from the initial population are as follows:

String 2: 0001010000100101

String 8: 0011001010010111

String 2: 0001010000100101

String 3: 0011111111111111

String 5: 1010011100000000

String 2: 0001010000100101

String 6: 0110110000111000

String 7: 0110000101101111

String 7: 0110000101101111

String 10: 0111001001001111

Step 4: Crossover

The strings in the mating pool are subjected to the crossover operation. In this operation, an exchange of properties between the pair of parent strings takes place from which a pair of offspring strings is produced. In our example, five mating pairs are formed from a population of 10 strings. It is noted that not all five mating pairs are subjected to crossover. For example, in a population of size 10, with $p_c = 0.8$, the expected number of mating pairs that are subjected to crossover is $0.8 \times 10/2$, or 4. In our example as shown in Table 9.3, it turns out that the fourth pair has not been subjected to crossover, and its strings are simply copied to the intermediate population.

Uniform crossover has been used in this example. Table 9.3 depicts the working principle of uniform crossover. Let us consider the case of the first mating pair as depicted in Table 9.3, where parent strings 2 and 8 participate in the crossover operation. While performing the tosses at each bit position, the outcomes turn out to be heads at position numbers 2, 3, 4, 7, 9, 10, 12, 13, 14, 15, and 16. Hence, swapping of the bits between the two parent strings 2 and 8 has taken place in these selected positions leading to the formation of two

TABLE 9.3

Crossover

Pair Number	Does Crossover Occur?		Uniform Crossover
1	Yes	Parent strings	0 0 0 1 0 1 0 0 0 0 1 0 0 1 0 1
			0 0 1 1 0 0 1 0 1 0 0 1 0 1 1 1
		Toss results	T H H H T T H T H H T H H H H H
		Offspring strings	0 0 1 1 0 1 1 0 1 0 1 1 0 1 1 1
			0 0 0 1 0 0 0 0 0 0 0 0 0 1 0 1
2	Yes	Parent strings	0 0 0 1 0 1 0 0 0 0 1 0 0 1 0 1
			0 0 1 1 1 1 1 1 1 1 1 1 1 1 1 1
		Toss results	H H T T H H H T H H H T H H H H T
		Offspring strings	0 0 0 1 1 1 0 1 1 1 1 1 1 1 1 1
			0 0 1 1 0 1 1 0 0 0 1 0 0 1 0 1
3	Yes	Parent strings	1 0 1 0 0 1 1 1 0 0 0 0 0 0 0 0
			0 0 0 1 0 1 0 0 0 0 1 0 0 1 0 1
		Toss results	T T H H H T H T H H H T T T H H
		Offspring strings	1 0 0 1 0 1 0 1 0 0 1 0 0 1 0 0 0 1
			0 0 1 0 0 1 1 0 0 0 0 0 0 1 0 0
4	No	Strings remain unchanged	0 1 1 0 1 1 0 0 0 0 1 1 1 0 0 0
			0 1 1 0 0 0 0 1 0 1 1 0 1 1 1 1
5	Yes	Parent strings	0 1 1 0 0 0 0 1 0 1 1 0 1 1 1 1
			0 1 1 1 0 0 1 0 0 1 0 0 1 1 1 1
		Toss results	H T H T T H H H T H H T T T H T
		Offspring strings	0 1 1 0 0 0 1 0 0 1 0 0 1 1 1 1
			0 1 1 1 0 0 0 1 0 1 1 0 1 1 1 1

offspring strings. In a particular bit position of the two parent strings, if both are either 1s or 0s, the swapping of bits does not make any change. Thus, in bit position numbers 2, 4, 10, 13, 14, and 16, there is no effect after swapping. Once the crossover operation for the first mating pair is over, the second mating pair (strings 2 and 3) is selected for the same. The rest of the mating pairs follow a similar process to that depicted in Table 9.3. It is noted that for the fourth pair, as the toss turns out to be "tails," no crossover operation has taken place, and the strings remain unchanged.

Step 5: Mutation

In this example, we considered the probability of mutation (p_m) as 0.005. Thus, in a population size of 10 and a string length of 16, the expected count of the mutated bit is about $0.005 \times 10 \times 16$ or 0.8. In our example, while performing the toss at every bit of the crossed-over pool, the outcome turns out to be true only in string number nine in its second bit position where there is 1. This position is marked by the box in Table 9.4. The mutation turns the value of 1 into 0 and a mutated pool is formed, which is shown in the second column of Table 9.4.

Step 6: Fitness Evaluation of the Mutated Pool

This step is similar to step 2. After estimating the decoded and real values of the substrings of the new population, that is, the mutated pool, the values of the objective function as well as the fitness function are evaluated as given in Table 9.5. With this step, the first generation of the genetic algorithm completes.

After the first generation, average and maximum values of fitness become 0.02847 and 0.044748, respectively, which gives a ratio (R) of 0.63622. Thus, it is noticed that the average fitness value has improved from 0.01888 to 0.02847,

TABLE 9.4

Mutation

Crossed-Over Pool	Mutated Pool
0011011010110111	0011011010110111
0001000000000101	0001000000000101
0001110111111111	0001110111111111
0011011000100101	0011011000100101
1001010100100001	1001010100100001
0010011000000100	0010011000000100
0110110000111000	0110110000111000
0110000101101111	0110000101101111
0⬛10001001001111	0010001001001111
0111000101101111	0111000101101111

TABLE 9.5

Fitness Evaluation of Mutated Pool

Mutated Pool							
Substring 1	Substring 2	D_1	D_2	x_1	x_2	$f(x_1,x_2)$	$F(x_1,x_2)$
00110110	10110111	54	183	2.1176	7.1765	96.789	0.010226
00010000	00000101	16	5	0.62745	0.19608	22.442	0.042658
00011101	11111111	29	255	1.1373	10	256.39	0.003885
00110110	00100101	54	37	2.1176	1.451	21.347	0.044748
10010101	00100001	149	33	5.8431	1.2941	96.85	0.01022
00100110	00000100	38	4	1.4902	0.15686	25.593	0.037604
01101100	00111000	108	56	4.2353	2.1961	37.472	0.025993
01100001	01101111	97	111	3.8039	4.3529	24.158	0.039748
00100010	01001111	34	79	1.3333	3.098	29.787	0.032481
01110001	01101111	113	111	4.4314	4.3529	25.93	0.037133

and the value of R has increased from 0.4497 to 0.63622, in one generation of the genetic algorithm. It is apparent from Table 9.5 that string number 4 gives the best fit in the new population, and the corresponding values of x_1 and x_2 are 2.1176 and 1.451, respectively. In Figure 9.10, the new population after the first generation (marked with the empty circles) and the best individual

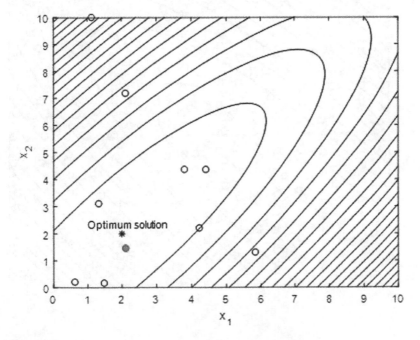

FIGURE 9.10
The new population and the best individual on the contour plot of the objective function.

(marked with the filled circle) are displayed on the contour plot. A comparison of Figures 9.9 and 9.10 reveals that after the first generation, there is a greater concentration of individual solutions at the vicinity of the optimum solution. Furthermore, after the first generation, the best individual is found to be closer to the optimum solution than that of the old generation.

Once the first generation is completed, we proceed to the next generation, which resumes from step 3. It is expected that in each generation, the values of average fitness and R are improved. Steps 3–6 are repeated until the stopping criterion is satisfied. The stopping criterion may be set based on either the maximum number of generations or a desired value of R. If the desired value of R is chosen as 0.99, then on average, 99% of the population converges to the maximum fitness value.

The MATLAB® coding for the aforesaid optimization problem given in Equation 9.2 using the genetic algorithm is presented in Section 9.5. The MATLAB® code was executed to solve the optimization problem. Figure 9.11 shows the spread of the initial population of size 1,000 in the search space of two variables x_1 and x_2. All 1,000 points in the population after 100 generations are displayed in Figure 9.12, from which it is manifested that the majority of the population has either reached the optimum solution or is clustered around the optimum solution. Figure 9.13 shows the plot between the number of

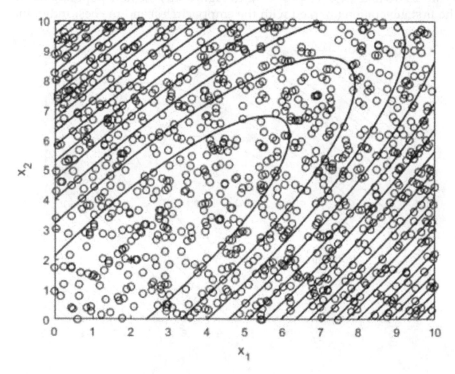

FIGURE 9.11
The spread of initial population of size 1,000 on the contour plot.

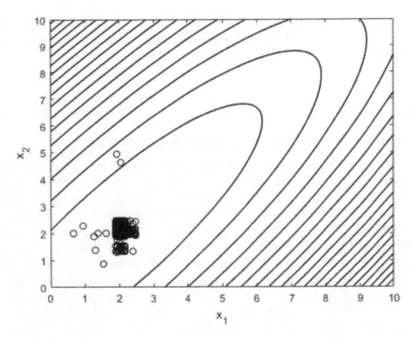

FIGURE 9.12
Scatter of population after 100 generation on the contour plot.

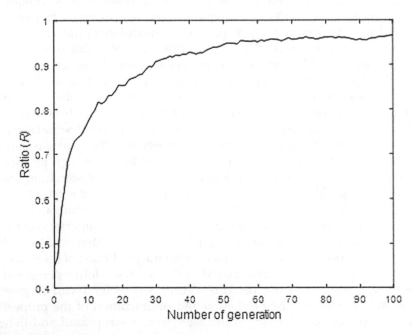

FIGURE 9.13
Number of generation versus *R*.

generations and the ratio (R). It is evident from Figure 9.13 that as the number of generation increases, more and more individuals of the population converge to the optimum fitness value. After 100 generations, R becomes 0.9652.

9.4 Application of Genetic Algorithm in Textiles

Blaga and Draghici (2005) optimized the knitting machine cam profile using the genetic algorithm. Sette et al. (1997) obtained set-point values and raw material characteristics for optimal quality of spun yarns using the genetic algorithm. In addition, they optimized the architecture of a neural network with the help of the genetic algorithm. Ghosh et al. (2005) used the genetic algorithm to obtain the best-fitted stress–strain curves of different spun yarns. In another work, Ghosh et al. (2013) studied the stress relaxation and creep behaviors of the core-spun worsted yarns using Eyring's model with the aid of the genetic algorithm. Lin (2003) developed a searching mechanism based on a genetic algorithm that can find several desired solutions of weaving parameters to produce woven fabrics within controlled costs. In addition, the system can simultaneously calculate the fractional cover of the fabric for each set of surveyed solutions in order to provide the designer with options for the functionality of this fabric. Amin et al. (2007) demonstrated a new technique for detecting the source of fault in spinning mills from spectrograms by using the genetic algorithm. Sette et al. (1996) proposed a model for optimization of the fiber-to-yarn process using the genetic algorithm. Das et al. (2013) developed a hybrid model based on an artificial neural network and the genetic algorithm for the selection of fiber properties to produce engineered yarn. They used the genetic algorithm to search for the best combination of fiber properties that can translate into reality a yarn with requisite quality by solving the constrained optimization problem deriving from the artificial neural network equations mapping the input-output relation between fiber-yarn properties. They concluded that the hybrid model captures both the high prediction power of an artificial neural network and the global solution searching ability of the genetic algorithm. Admuthe and Apte (2009) developed a hybrid neuro-genetic model for a ring yarn spinning process and cost optimization. They reported that the performance of the hybrid model was superior compared to current manual machine intervention. Lv et al. (2011) attempted to apply a genetic algorithm–support vector machine for the prediction of the spinning quality of yarn. Subramanian et al. (2007) studied the relative performance of the backpropagation neural network algorithm combined with the genetic algorithm approach for the prediction and optimization of the properties of yarn produced on a jet ring spinning system. Guruprasad and Behera (2010) explored the possibility of using artificial neural network and genetic algorithm approaches to predict the overall bending rigidity of woven fabrics.

They developed an artificial neural network model based on backpropagation learning and assessed its prediction performance. They reported that the backpropagation algorithm complements the genetic algorithm and performs a local search efficiently to reach global optima. Huang and Tang (2006a) proposed a systematic approach, which is the application of the Taguchi method, neural network, and genetic algorithm, for determining parameter values in melt spinning processes to yield optimal qualities of denier and tenacity in as-spun fibers. The results showed that the artificial neural network–genetic algorithm approach could yield better qualities of denier and tenacity in polypropylene as-spun fibers than the Taguchi method. In another work, Huang and Tang (2006b) proposed a quantitative procedure for determining the values of critical process parameters in melt spinning to optimize denier, tenacity, breaking elongation, and denier variance in as-spun polypropylene yarn. Banerjee et al. (2013) developed a hybrid model of cotton yarn strength by using a fuzzy expert system and the genetic algorithm. They used the genetic algorithm for better tuning of membership functions of the fuzzy expert system. They concluded that the genetic-fuzzy expert system has significantly better prediction accuracy and consistency than that of the fuzzy expert system alone. Pan et al. (2011) proposed a novel method based on the genetic algorithm to recognize the layout of color yarns of yarn-dyed fabric from the color pattern. Experiments on some color patterns recognized from actual yarn-dyed fabrics, some color patterns simulated manually, and some color patterns including error color information of floats prove that the method is effective for detecting the layout of color yarns from the color pattern of yarn-dyed fabric, and it has a fault-tolerance ability to some degree.

9.4.1 Application of Genetic Algorithm in Fitting Stress–Strain Curve of Fibers

Ghosh et al. (2015) used the genetic algorithm for the purpose of fitting the stress–strain relationship derived from the nonlinear viscoelastic model of Eyring (1936) on the experimental data of polyester and viscose filaments. The most attractive feature of Eyring's model is that it offers the possible identification of the molecular mechanism and, hence, helps in unraveling some aspects of the structure dependence of mechanical behavior. In addition, it provides a common basis to explain the strain-strain curve, stress relaxation, and creep behaviors of fibrous materials. Although Eyring's model was developed in the 1930s, its applications have been limited to textile materials due to the mathematical rigors involved in the computational works. Even in a few reported works, many assumptions are made to simplify the equations derived from Eyring's model. With the advent of very high computational speed and mathematical techniques such as the genetic algorithm, it has been possible to solve this problem using the original equations. A brief description of obtaining the stress–strain relationship from Eyring's nonlinear viscoelastic model is provided.

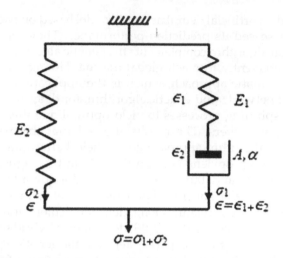

FIGURE 9.14
Eyring's three-elements nonlinear viscoelastic model.

Eyring's three-elements nonlinear viscoelastic model is shown in Figure 9.14. For the spring in the right-hand arm of the model, the stress–strain relationship is given by

$$\varepsilon_1 = \frac{\sigma_1}{E_1} \tag{9.4}$$

where $\varepsilon_1, \sigma_1,$ and E_1 are the strain, stress, and modulus of the spring, respectively. By differentiating Equation 9.4 with respect to time t, we get

$$\frac{d\varepsilon_1}{dt} = \frac{1}{E_1}\frac{d\sigma_1}{dt} \tag{9.5}$$

For the dashpot, the strain rate of the non-Newtonian fluid is represented by the hyperbolic-sine law of viscous flow:

$$\frac{d\varepsilon_2}{dt} = A\,\sinh\alpha\sigma_1 \tag{9.6}$$

where ε_2 and σ_1 are the strain and stress of the dashpot, respectively, and σ_1 and α, are the two constants of the non-Newtonian fluid. Constants A and α are the indirect measures of activation free energy and activation volume of the flow, respectively. As the spring and dashpot are in series, the total strain of the right arm is given by

$$\varepsilon = \varepsilon_1 + \varepsilon_2 \tag{9.7}$$

By differentiating Equation 9.7, we have

$$\frac{d\varepsilon}{dt} = \frac{d\varepsilon_1}{dt} + \frac{d\varepsilon_2}{dt} \tag{9.8}$$

Substituting the relations from Equations 9.5 and 9.6, Equation 9.8 becomes

$$\frac{d\varepsilon}{dt} = \frac{1}{E_1}\frac{d\sigma_1}{dt} + A \sinh \alpha \sigma_1 \tag{9.9}$$

For the spring in the left-hand arm of the model, the stress–strain relationship is given by

$$\varepsilon = \frac{\sigma_2}{E_2} \tag{9.10}$$

where ε, σ_2, and E_2 are the strain, stress, and modulus of the left-hand spring, respectively. By differentiating Equation 9.10, we have

$$\frac{d\varepsilon}{dt} = \frac{1}{E_2}\frac{d\sigma_2}{dt} \tag{9.11}$$

As the right and left arms of the model are in parallel, the total stress σ can be expressed as

$$\sigma = \sigma_1 + \sigma_2 \tag{9.12}$$

By differentiating Equation 9.12, we have

$$\frac{d\sigma}{dt} = \frac{d\sigma_1}{dt} + \frac{d\sigma_2}{dt} \tag{9.13}$$

Eliminating the relations from Equations 9.9 through 9.12, Equation 9.13 becomes

$$\frac{d\sigma}{dt} = \frac{d\varepsilon}{dt}(E_1 + E_2) - E_1 A \sinh(\sigma - E_2\varepsilon)\alpha \tag{9.14}$$

For a constant rate of strain, $d\varepsilon/dt = r$, where $r = $ constant. At $t = 0$, $\varepsilon = 0$, and hence, $\varepsilon = rt$.

In order to solve Equation 9.14, we put

$$z = \sinh(\sigma - E_2\varepsilon)\alpha \tag{9.15}$$

By differentiating Equation 9.15, we get

$$\frac{dz}{dt} = \cosh \alpha (\sigma - \varepsilon E_2) \frac{d}{dt} \alpha (\sigma - \varepsilon E_2) \qquad (9.16)$$

After further simplification, Equation 9.16 becomes

$$\frac{dz}{dt} = E_1 \alpha \sqrt{1+z^2}\,(r - Az) \qquad (9.17)$$

The solution of Equation 9.17 is found to be

$$z = \tan\left[2\tan^{-1}\left(e^{\sqrt{r^2+A^2}\,(E_1\alpha t + c)}\right) - \frac{\pi}{2} - \cot^{-1}\frac{r}{A}\right] \qquad (9.18)$$

where

$$c = \frac{1}{\sqrt{r^2+A^2}} \ln\left| \tan\left(\frac{\pi}{4} + \frac{1}{2}\right) \cot^{-1}\frac{r}{A} \right|$$

Substituting, $z = \sin h\alpha(\sigma - \varepsilon E_2)$ and $t = \varepsilon/r$ in Equation 9.18, we get the stress–strain relation as

$$\sigma = \frac{1}{\alpha}\sinh^{-1}\left\{\tan\left[2\tan^{-1}\left(e^{\sqrt{r^2+A^2}\left(\frac{E_1\alpha\varepsilon}{r}+c\right)}\right) - \frac{\pi}{2} - \cot^{-1}\frac{r}{A}\right]\right\} + \varepsilon E_2 \qquad (9.19)$$

Equation 9.19 can be fitted on the experimental stress–strain curves of polyester and viscose fibers using the least square method—that is, by minimizing the sum of squares of difference between the experimental values of stress ($\sigma_{\exp}$) and the predicted values of the stress (σ) at different levels of strain. Thus, the objective function of the optimization problem becomes

Minimize $f(E_1, E_2, A, \alpha)$

$$= \sum \left(\sigma_{\exp} - \frac{1}{\alpha}\sinh^{-1}\left\{\tan\left[2\tan^{-1}\left(e^{\sqrt{r^2+A^2}\left(\frac{E_1\alpha\varepsilon}{r}+c\right)}\right) - \frac{\pi}{2} - \cot^{-1}\frac{r}{A}\right]\right\} - \varepsilon E_2 \right)^2 \qquad (9.20)$$

The exotic nature of Equation 9.20 makes it difficult to obtain the best-fitted curve on the experimental stress–strain data by means of the classical optimization method. A nontraditional search-based optimization technique such as the genetic algorithm is an appropriate method to solve such a type of complex problem.

For experimental purposes, high-tenacity polyester multifilament yarn of 210 denier consisting of 72 monofilaments and viscose multifilament yarn of 150 denier consisting of 40 monofilaments were considered. The yarn samples were conditioned for 24 hours at the standard atmospheric condition of 65% relative humidity and 27°C before the experiments. To study the stress–strain curves of yarns, tensile tests were performed in an Instron tensile tester at a constant strain rate (r) of 1 min^{-1} and gauge length of 500 mm. For each type of yarn, 50 tests were conducted. A typical stress–strain curve having tenacity and breaking strain close to the average values was selected for curve fitting.

Using genetic algorithms, the optimization problem of Equation 9.20 was solved, and the optimum values of E_1, E_2, A, and α were determined for both polyester and viscose filaments. The maximum number of generation as well as population size were set to 1,000. The probability of crossover (p_c) and probability of mutation (p_m) were set to 0.7 and 0.001, respectively. Table 9.6 shows the optimum values of these constants for polyester and viscose filaments.

The experimental and fitted stress–strain curves obtained with Equation 9.19 for polyester and viscose filaments are depicted in Figure 9.15. The experimental curves are shown by the solid lines, and the corresponding fitted curves are shown by the dotted lines. Invariably for both filaments, a high degree of coefficient of determination (R^2) justifies a good fit to the experimental data.

It is evident from Table 9.6 that the values of E_1, E_2, and A are higher for polyester yarn than for viscose yarn, whereas an opposite trend is seen for α. The higher values of E_1 and E_2 for polyester yarn can be attributed to its higher crystallinity, modulus, and elasticity compared to viscose yarn. Constants A and α correspond to the activation free energy and activation volume of the flow, respectively, for the non-Newtonian dashpot of Eyring's model. A stiff dashpot shows a higher value of A and a lower value of α in comparison to a slack dashpot. As the viscous part dominates in viscose yarn, it shows a lower value of A and a higher value of α in comparison to the polyester

TABLE 9.6

Constants of Eyring's Model for Polyester and Viscose Yarns

Constants	Polyester	Viscose
E_1 (gpd)	106.01	75.73
E_2 (gpd)	45.74	5.07
A (s^{-1})	5.28×10^{-4}	6.19×10^{-6}
α (gpd^{-1})	5.19	17.90

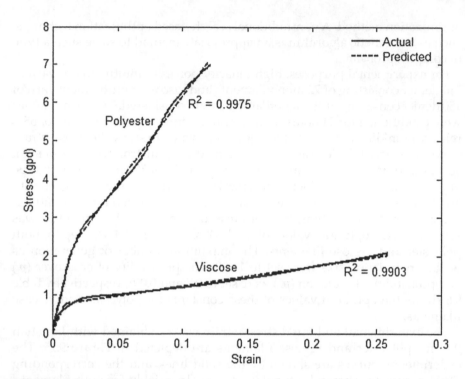

FIGURE 9.15
Experimental and fitted stress–strain curves for polyester and viscose yarns.

yarn. Higher activation volume of flow for viscose yarn is accountable to the greater plastic flow in its stress–strain curve than that of polyester yarn. As a consequence of that, viscose yarn has low modulus compared to polyester yarn. The advent of a nontraditional search-based optimization technique such as the genetic algorithm makes it easy to solve the complex curve-fitting problem with Eyring's equations.

9.5 MATLAB® Coding

MATLAB® Coding for the Optimization Problem Given in Equation 9.2 Using the Genetic Algorithm

```
clear all
close all
clc
format short g
%Step 1- Generation of initial population
pop_size=1000;
maxgen=1000;
```

```
pc=0.8;
pm=0.005;
vlb=[0 0];
vub=[10 10];
bits=[8 8];
n=length(bits);
len_chrom=sum(bits);
old_gen=round(rand(pop_size,len_chrom));
for i=1:n
    accuracy(i)=(vub(i)-vlb(i))./(2^bits(i)-1);
end
gen=0;
Fx(1:pop_size)=0;
%Step 2 - Fitness evaluation
dec_val(1:pop_size,1:n)=0;
m=1;
for i=1:n
    cg=m;
    for j=bits(i)-1:-1:0
        dec_val(:,i)=dec_val(:,i)+(2^j*old_gen(:,cg));
        cg=cg+1;
        m=cg;
    end
end
for j=1:n
    for i=1:pop_size
        x(i,j)=vlb(j)+dec_val(i,j).*accuracy(j);
    end
end
for i=1:pop_size
        f(i)=4*x(i,1).^2 + 3*x(i,2).^2 - 6*x(i,1).*x(i,2) - 4*x(i,1)+24;
        Fx(i)=1/(1+f(i));
end
Fitness=[f' Fx'];
ratio=mean(Fx)/max(Fx);
%Step 3 - Reproduction (Roulette-wheel selection)
 while ratio<=0.99
      C(1:pop_size)=0;
        for i=1:pop_size
            A(i)=Fx(i)./mean(Fx);
            B(i)=A(i)./pop_size;
        end
        CC=0;
        for i=1:pop_size
            CC= CC+B(i);
            C(i)=CC;
        end
        Col_A=A';
        Col_B=B';
        Col_C=C';
        D=rand(pop_size,1);
        co=1;
        F(1:pop_size)=0;
        for i=1:pop_size
            for j=1:pop_size
                if D(i)<C(j)
```

```
                    mating_pool(co,:)=old_gen(j,:);
                    co=co+1;
                    F(j)=F(j)+1;
                    E(i)=j;
                    break
                    end
                end
            end
%Step 4: Cross Over
    for i=1:2:pop_size
        if ceil(pop_size*rand/2)<=ceil(pop_size*pc/2)
            for j=1:len_chrom
                if rand<=0.5
                    crossed_over_pool(i,j)=mating_pool(i,j);
                    crossed_over_pool(i+1,j)=mating_pool(i+1,j);
                else
                    crossed_over_pool(i,j)=mating_pool(i+1,j);
                    crossed_over_pool(i+1,j)=mating_pool(i,j);
                end
            end
            G(i)='Y';
            G(i+1)='Y';
        else
            crossed_over_pool(i,:)=mating_pool(i,:);
            crossed_over_pool(i+1,:)=mating_pool(i+1,:);
            G(i)='N';
            G(i+1)='N';
        end
    end
%Step 5: Mutation
    mu=0;
    for i=1:pop_size
        for j=1:len_chrom
            if ceil(pop_size*len_chrom*rand)<=ceil(pop_size*len_chrom*pm)
                if crossed_over_pool(i,j)==0
                    mutated_pool(i,j)=1;
                else
                    mutated_pool(i,j)=0;
                end
                mu=mu+1;
                [i j];
            else
                mutated_pool(i,j)=crossed_over_pool(i,j);
            end
        end
    end
%Step 6 - Fitness evaluation of the mutated pool
    old_gen=mutated_pool;
    dec_val(1:pop_size,1:n)=0;
    m=1;
    for i=1:n
        cg=m;
        for j=bits(i)-1:-1:0
            dec_val(:,i)=dec_val(:,i)+(2^j*old_gen(:,cg));
            cg=cg+1;
            m=cg;
```

```
      end
   end
   for j=1:n
      for i=1:pop_size
         x(i,j)=vlb(j)+dec_val(i,j).*accuracy(j);
      end
   end
   for i=1:pop_size
      f(i)=4*x(i,1).^2 + 3*x(i,2).^2 - 6*x(i,1).*x(i,2) - 4*x(i,1)+24;
      Fx(i)=1/(1+f(i));
   end
   for i=1:pop_size
      if Fx(i)==max(Fx)
         best_fit=i;
      break;
      end
   end
   gen=gen+1
   if gen==maxgen
      break
   end
   ratio=mean(Fx)/max(Fx);
end
%Best Values of variables and objective function
x(best_fit,:)
eval_fit=4*x(best_fit,1).^2 + 3*x(best_fit,2).^2 - 6*x(best_
fit,1).*x(best_fit,2) - 4*x(best_fit,1)+24
```

9.6 Summary

This chapter presents an overview of the different steps of a genetic algorithm. A worked-out problem shows every step of the genetic algorithm along with its coding in MATLAB® language. A survey of literature on application of the genetic algorithm in various textile problems is discussed. This chapter also includes the results of a study on application of the genetic algorithm in a complex curve-fitting problem reported by the author.

References

Admuthe, L. S. and Apte, S. D. 2009. Neuro-genetic cost optimization model: Application of textile spinning process. *International Journal of Computer Theory and Engineering*, 1(4), 441–444.

Amin, A. E., Geheni, A. S., Hawary, I. A. and Beali, R. A. 2007. Detecting the fault from spectrograms by using genetic algorithm techniques. *AUTEX Research Journal*, 7(2), 80–88.

Banerjee, D., Ghosh, A. and Das, S. 2013. Yarn strength modelling using genetic fuzzy expert system. *Journal of Institute of Engineers (India): Series E*, 93(2), 83–90.

Blaga, M. and Draghici, M. 2005. Application of genetic algorithm in knitting technology. *Journal of Textile Institute*, 96(3), 175–178.

Das, S., Ghosh, A., Majumdar, A. and Banerjee, D. 2013. Yarn engineering using hybrid artificial neural network and genetic algorithm model. *Fibers and Polymers*, 14(7), 1220–1226.

Davis, L. 1991. *Handbook of Genetic Algorithms*, Van Nostrand, Reinhold, New York.

Deb, K. 2005. *Optimization for Engineering Design: Algorithms and Examples*, Prentice Hall, New Delhi, India.

Eyring, H. 1936. Viscosity, plasticity, and diffusion as examples of absolute reaction rates. *Journal of Chemical Physics*, 4, 283–291.

Ghosh, A., Das, S. and Banerjee, D. 2013. Simulation of yarn stress relaxation and creep behaviors using genetic algorithm. *Indian Journal of Fibres and Textile Research*, 38(4), 375–379.

Ghosh, A., Das, S. and Saha, B. 2015. Simulation of stress-strain curves of polyester and viscose filaments. *Journal of Institute of Engineers (India): Series E*, 96(2), 139–143.

Ghosh, A., Ishtiaque, S. M. and Rengasamy, R. S. 2005. Stress-strain characteristics of different spun yarns as a function of strain rate and gauge length. *Journal of Textile Institute*, 96(2), 99–104.

Goldberg, D. E. 1989. *Genetic Algorithms in Search, Optimization and Machine Learning*, Addison-Wesley, New York.

Guruprasad, R. and Behera, B. K. 2010. A prediction of bending of woven fabrics by soft computing. *Seventh International Conference on Textile Science*, Liberec, Czech Republic.

Holland, J. H. 1975. *Adaptation in Natural and Artificial System*, University of Michigan Press, Ann Arbor, MI.

Huang, C. C. and Tang, T. T. 2006a. Optimizing multiple qualities in as-spun polypropylene yarn by neural networks and genetic algorithms. *Journal of Applied Polymer Science*, 100, 2532–2541.

Huang, C. C. and Tang, T. T. 2006b. Parameter optimization in melt spinning by neural networks and genetic algorithms. *International Journal of Advanced Manufacturing Technology*, 27, 1113–1118.

Lin, J. J. 2003. A genetic algorithm for searching the weaving parameters of woven fabrics. *Textile Research Journal*, 73(2), 105–112.

Lv, Z. J., Xiang, Q. and Yang, J. G. 2011. Application of genetic algorithm-support vector machine for prediction of spinning quality. *Proceedings of the World Congress on Engineering*, II, London.

Pan, R., Gao, W., Liu, J. and Wang, H. 2011. Genetic algorithm-based detection of the layout of color yarns. *Journal of the Textile Institute*, 102(2), 172–179.

Sette, S., Boullart, L., Langenhove, V. L. and Kiekens, P. 1997. Optimizing the fibre-to-yarn production process with a neural network/genetic algorithm approach. *Textile Research Journal*, 67(2), 84–92.

Sette, S., Langenhove, V. L. and Boullart, L. 1996. Optimisation of the fibre-to-yarn process using genetic algorithms. *Twenty-third International Cotton Conference Bremen: Optimisation Production Process*, Germany, March 6–9, 233–246.

Subramanian, S. N., Venkatachalam, A. and Subramaniam, V. 2007. Prediction and optimization of yarn properties using genetic algorithm/artificial neural network. *Indian Journal of Fibre and Textile Research*, 32, 409–413.

10

Particle Swarm Optimization

10.1 Introduction

Particle swarm optimization (PSO) is a relatively recent heuristic search method whose mechanics are inspired by the social behavior of bird flocking or fish schooling. This algorithm was developed by Kennedy and Eberhart in 1995. It is similar to the genetic algorithm in the sense that these two evolutionary heuristics are population-based search methods.

Suppose that a group of flying birds is searching for food in an area where there is only one piece of food. The central idea of the PSO algorithm is derived from the observation of the collective movement of birds in search of food. Each bird in the group has its own position and velocity at any given moment. In order to search for food, a bird keeps on changing its position by adjusting its velocity. A bird adjusts its velocity by combining its self-experiences with social experiences. This food-searching process by a group of birds is artificially simulated in PSO for solving nonlinear optimization problems; thus, it becomes a population-based stochastic optimization technique. A bird would be termed a *particle*.

PSO uses a number of particles that constitute a swarm flying within the search space looking for the best solution. Thus, PSO is initiated with a population of solutions where each solution is considered as the position of a particle in the search space. All of the particles in a swarm have fitness values that can be estimated from the objective function. Each particle in the swarm preserves its individual best performance; furthermore, it keeps track of the best performance of the group. Each particle continually modifies its position by adjusting the velocity according to its own best performance as well as the best performance of the group. Eventually, most of the particles converge to the optimum solution.

10.2 Particle Swarm Optimization

PSO is a population-based stochastic optimization technique inspired by the social behavior of bird flocking. There are many similarities between

PSO and genetic algorithms. Both methods are initialized with a population of random solutions, and the searches for optimal solutions are proceeded by updating the population in the next iterations. However, unlike the genetic algorithm, PSO has no evolutionary operators, such as crossover and mutation. PSO simulates the choreography of bird swarm that flies through the problem space, with each bird adjusting its traveling speed, dynamically corresponding to the flying experiences of itself and its colleagues (Kennedy and Eberhart, 1995).

PSO is initialized by a population of random solutions, and each potential solution is assigned a randomized velocity. The potential solutions, called *particles*, are then flown through the problem space. The fitness function is evaluated for each particle. Each particle keeps track of its coordinates in the problem space, which are associated with the best solution or fitness value achieved so far. This value is called *personal best* (pbest). Another best value that is tracked by the global version of the particle swarm optimizer is the overall best value and its location, obtained so far by any particle in the population. This value is termed *global best* (gbest). Suppose that the location and velocity of the ith particle of the swarm at a time step t are represented by N-dimensional vectors $X_i^{(t)} = (x_{i1}, x_{i2}, \ldots, x_{iN})$ and $V_i^{(t)} = (v_{i1}, v_{i2}, \ldots, v_{iN})$, respectively. If the best positions experienced so far by the ith particle and the whole swarm are denoted as $pbest_i^{(t)} = (p_{i1}, p_{i2}, \ldots, p_{iN})$ and $gbest_i^{(t)}$, respectively, then in the next time step $t + 1$, the ith particle of the swarm updates its velocity as well as location as follows (Shi and Eberhart, 1998a,b):

$$V_i^{(t+1)} = wV_i^{(t)} + C_1 r_1 (pbest_i^{(t)} - X_i^{(t)}) + C_2 r_2 (gbest^{(t)} - X_i^{(t)}) \tag{10.1}$$

$$X_i^{(t+1)} = X_i^{(t)} + V_i^{(t+1)} \tag{10.2}$$

where $i = 1, 2, \ldots, S$ (size of the swarm); w is the inertia weight; C_1 and C_2 are the learning factors; and r_1 and r_2 are the random numbers uniformly distributed in the range (0, 1).

According to Equations 10.1 and 10.2, the next displacement of a particle depends on three fundamental elements, such as inertia effect (its own velocity), personal influence (its best performance), and social influence (best performance of the group). In Equation 10.1, the first term $wV_i^{(t)}$ is the inertia effect, responsible for keeping the particle moving in the same direction it was originally heading. The second term $C_1 r_1 (pbest_i^{(t)} - X_i^{(t)})$ is the personal influence, accountable for the local search, and it causes the particle to return to the best position it has experienced so far. The third term $C_2 r_2 (gbest^{(t)} - X_i^{(t)})$ is the social influence, liable for global search by causing the particle to move to the best region the swarm has found so far. A schematic movement of a particle is depicted in Figure 10.1. It is evident from Figure 10.1 that the next position $X_i^{(t+1)}$ of a particle is determined by its current position $X_i^{(t)}$, its current velocity $V_i^{(t)}$, its own best performance achieved so far $pbest_i^{(t)}$, and the best performance received from its neighbors so far $gbest_i^{(t)}$.

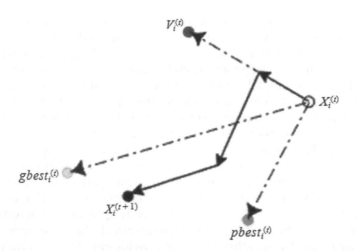

FIGURE 10.1
Schematic movement of a particle based on Equations 10.1 and 10.2.

It is essential to choose optimum values for the parameters for the best performance of PSO for different types of applications. So, the selections of the important parameters, such as inertia weight (w), learning factors (C_1, C_2), *pbest*, and *gbest*, have to be taken care of, which are discussed as follows.

The inertia weight w is utilized to adjust the influence of the previous velocity on the current velocity and to balance between global and local exploration abilities of the flying particle (Umapathy et al., 2010). A larger inertial weight implies stronger global exploration ability (i.e., searching new areas), which will permit the particle to escape from a local minimum. A smaller inertial weight leads to stronger local exploration ability, which confines the particle searching within a local range near its present position and thus helps convergence. Hence, proper selection of the inertial weight can provide a balance between global and local exploration abilities to reduce the average number of iterations. Shi and Eberhart (1998a,b) put forward the concept of linearly decreasing inertial weight for a better search as follows:

$$w = (w_{max} - w_{min}) \times \frac{iter_{max} - iter}{iter_{max}} + w_{min} \tag{10.3}$$

where $iter_{max}$ is the maximum iterations, $iter$ is the current iteration number, and w_{max} and w_{min} are the initial and final values of w, respectively.

A larger value of inertial weight during the initial iterations promotes global exploration of the search space, and a gradual decline of the inertial weight with the number of iterations promotes local exploration of the search space. Generally, it has been found that the values of 1.2 and 0.2 for w_{max} and w_{min}, respectively, can accelerate the convergence rate. Thus, an initial value of around 1.2 and a gradual decline toward 0.2 can be considered good choices for w.

The constants C_1 and C_2 are the two learning factors that represent the weighting of the stochastic acceleration terms that pull each particle toward the *pbest* and *gbest* positions. Thus, C_1 and C_2 are the balance factors between the effect of self-knowledge and social knowledge in moving the particle toward the target. The adjustments of these constants change the amount of tension in the system. Low values allow particles to roam far from target regions before being tugged back, while high values result in abrupt movements toward or past the target regions. The constants C_1 and C_2 are also termed as *cognitive coefficient* and *social coefficient*, respectively. The cognitive parameter represents the tendency of individuals to duplicate past behaviors that have proven successful, whereas the social parameter represents the tendency to follow the successes of others. Generally, C_1 and C_2 are set to 2.0, which will make the search cover all surrounding regions centered on *pbest* and *gbest*. Also, if the learning factors are identical, the same importance is given to local (nearby) searching and global (wide-ranging) searching; hence, both parts contribute equally to the success of particle swarm searching.

The *pbest* is the best position of the particle attained so far and can be considered as the particle's memory; one memory slot is allotted to each particle. The best location does not necessarily always depend on the value of the fitness function. To adapt to different problems, many constraints can be applied to the definition of the best location. In certain nonlinear constrained optimization problems, the particles remember the positions in feasible space and disregard unfeasible solutions. This simple alteration successfully locates the optimum solution for a series of benchmark problems.

The *gbest* is the best position that the neighbors of a particle have achieved so far. It acts like an attractor, pulling all the particles toward it. In the long run, all particles will converge to this position; hence, if it is not updated on a regular basis, the swarm may converge prematurely.

10.2.1 Flowchart of Particle Swarm Optimization

Step 1: Initialization

Initialize swarm size, maximum number of iterations, number of variables, and lower and upper boundaries of variables. Set the values of w, C_1, and C_2. Generate initial positions of the swarm X^0, which are uniformly distributed in the search space. Set $t = 0$.

Step 2: Fitness Evaluation

Evaluate each particle's position according to the objective function.

Step 3: Updating pbest

If a particle's current position is better than its previous best position, update it. We compare each particle's current position with its previous best position in terms of fitness value, and if the current position of a particle is found to be better than its previous best position, we assign the current position as *pbest*, or else its previous best position will remain as *pbest*.

Step 4: Updating gbest

If the current population's overall best position is better than the population's overall previous best position, update it to *gbest*.

Step 5: Updating Velocity of Each Particle

In this step, we update the velocity of each particle using Equation 10.1.

Step 6: Updating Position of Each Particle

We move each particle to its new position according to Equation 10.2.

If the desired criterion is not met, set $t = t + 1$ and go to step 2; otherwise, stop the process. When the process is stopped, the optimum solution is given by *gbest*. A flowchart of the PSO algorithm is illustrated in Figure 10.2.

10.3 Step-by-Step Working Principle of Particle Swarm Optimization

Let us consider the following optimization problem with Beale's function to explain the working principle of the PSO:

$$\text{Minimize} \quad f(x_1, x_2) = (1.5 - x_1 + x_1 x_2)^2$$
$$+ (2.25 - x_1 + x_1 x_2^2)^2 + (2.625 - x_1 + x_1 x_2^3)^2 \qquad (10.4)$$

Subject to $-5 \le x_1, x_2 \le 5$, where x_1 and x_2 are the real variables.

Surface and contour plots of Beale's function are displayed in Figures 10.3 and 10.4, respectively. It can be shown that the minimum value of the function is 0, which is obtained at $x_1 = 3$ and $x_2 = 0.5$. In Figures 10.3 and 10.4, the optimum solution is marked by an asterisk (*).

Various steps involved to solve the optimization problem of Equation 10.4 using PSO are discussed as follows for the first two iterations.

Step 1: Initialization

The typical value of swarm size N is around 1,000; however, for the sake of simplicity, here we have considered $N = 6$. Since there are two variables

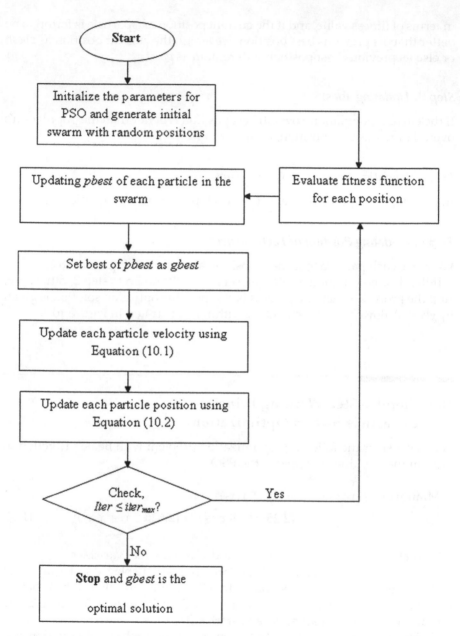

FIGURE 10.2
Flowchart of PSO algorithm.

x_1 and x_2 in the optimization problem of Equation 10.4, it becomes a two-dimensional search space, which is bounded by $-5 \leq x_1, x_2 \leq 5$. Suppose that within the given boundaries of search space, the swarm has the following initial positions:

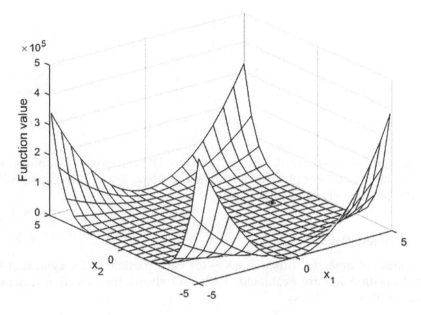

FIGURE 10.3
Surface plot of Beale's function.

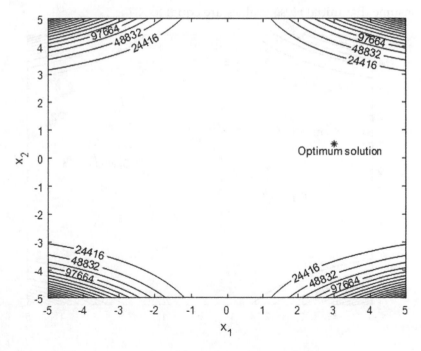

FIGURE 10.4
Contour plot of Beale's function.

$$X^{(0)} = \begin{bmatrix} 1 & -3 \\ 4 & 2 \\ -3 & -1 \\ -4 & 3 \\ -1 & 4 \\ 3 & 1 \end{bmatrix}$$

Figure 10.5 shows the initial swarm position. The values of w, C_1, and C_2 and the maximum number of iterations are set to 0.5, 2, 2, and 100, respectively. We also set the iteration counter $t = 0$.

Step 2: Fitness Evaluation

The values of objective function $f(X^{(0)})$ for each particle in the swarm at its initial position $X^{(0)}$ are evaluated. Table 10.1 shows the objective function values of all six particles.

Step 3: Updating pbest

We assume that initial $pbest^{(0)}$ values are equal to $X^{(0)}$.

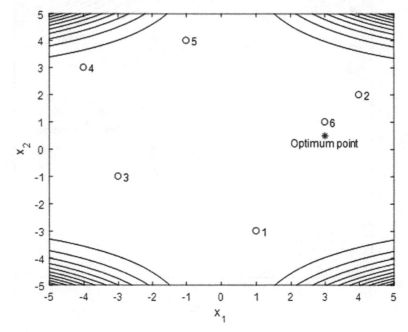

FIGURE 10.5
Initial swarm position.

TABLE 10.1

Fitness Values of Swarm at Initial Positions

Particle Number	$X^{(0)}$	$f(X^{(0)})$	$pbest^{(0)}$	$gbest^{(0)}$
1	[1 −3]	755.2	[1 −3]	[3 1]
2	[4 2]	1171.2	[4 2]	[3 1]
3	[−3 −1]	135.7	[−3 −1]	[3 1]
4	[−4 3]	11,204	[−4 3]	[3 1]
5	[−1 4]	3,810	[−1 4]	[3 1]
6	[3 1]	14.203	[3 1]	[3 1]

Step 4: Updating gbest

As the optimization problem of Equation 10.4 is a minimization one, it is evident from Table 10.1 that particle number 6 at its initial position [3 1] gives the overall best value. Figure 10.5 also illustrates that particle number 6 is located closer to the optimum solution than the other particles. Thus, the initial $gbest^{(0)}$ is [3 1].

Step 5: Updating Velocity of Each Particle

We set the initial velocity of swarm $V^{(0)}$ equal to $X^{(0)}$. As the optimization problem has two variables, both random numbers r_1 and r_2 will also have two dimensions for each particle. Table 10.2 shows the randomly generated values of r_1 and r_2 for all six particles. The velocity of each particle is updated according to Equation 10.1. For an example, the updated velocity of the first particle is estimated as follows:

$$V_1^{(1)} = wV_1^{(0)} + C_1r_1(pbest_1^{(0)} - X_1^{(0)}) + C_2r_2(gbest^{(0)} - X_1^{(0)})$$

$$= \begin{bmatrix} 0.5 \times 1 \\ 0.5 \times (-3) \end{bmatrix}^T + \begin{bmatrix} 2 \times 0.42573 \times (1-1) \\ 2 \times 0.64762 \times (-3+3) \end{bmatrix}^T + \begin{bmatrix} 2 \times 0.64444 \times (3-1) \\ 2 \times 0.67902 \times (1+3) \end{bmatrix}^T$$

$$= \begin{bmatrix} 0.5 \\ -1.5 \end{bmatrix}^T + \begin{bmatrix} 2 \times 0.42573 \times 0 \\ 2 \times 0.64762 \times 0 \end{bmatrix}^T + \begin{bmatrix} 2 \times 0.64444 \times 2 \\ 2 \times 0.67902 \times 4 \end{bmatrix}^T$$

$$= \begin{bmatrix} 3.0778 \\ 3.93213 \end{bmatrix}^T$$

$$= [3.0778 \quad 3.93213]$$

Similarly, other particles in the swarm update their velocities. The updated velocities of the whole swarm are given in Table 10.2.

TABLE 10.2

Initial and Updated Velocities of Swarm in First Iteration

Particle Number	$V^{(0)}$	r_1	r_2	$V^{(1)}$
1	[1 −3]	[0.42573 0.64762]	[0.64444 0.67902]	[3.0778 3.93213]
2	[4 2]	[0.63579 0.20893]	[0.94517 0.70928]	[0.10965 −0.41856]
3	[−3 −1]	[0.23623 0.6073]	[0.1194 0.45014]	[−0.06725 1.3006]
4	[−4 3]	[0.45873 0.77029]	[0.66194 0.35022]	[7.2672 0.099128]
5	[−1 4]	[0.66201 0.84193]	[0.41616 0.83292]	[2.8293 −2.9975]
6	[3 1]	[0.25644 0.58225]	[0.61346 0.54074]	[1.5 0.5]

TABLE 10.3

New Position of Swarm after First Iteration

Particle Number	$X^{(0)}$	$V^{(1)}$	$X^{(1)}$
1	[1 −3]	[3.0778 3.93213]	[4.0778 0.93213]
2	[4 2]	[0.10965 −0.41856]	[4.1097 1.5814]
3	[−3 −1]	[−0.06725 1.3006]	[−3.0672 0.30055]
4	[−4 3]	[7.2672 0.099128]	[3.2672 3.0991]
5	[−1 4]	[2.8293 −2.9975]	[1.8293 1.0025]
6	[3 1]	[1.5 0.5]	[4.5 1.5]

Step 6: Updating Positions of Each Particle

Each particle in the swarm will fly to its new position according to Equation 10.2. As an example, the new position of the first particle can be computed as follows:

$$X_1^{(1)} = X_1^{(0)} + V_1^{(1)}$$
$$= [1 \quad -3] + [3.0778 \quad 3.93213]$$
$$= [4.0778 \quad 0.93213]$$

Table 10.3 shows the new positions of all six particles. We increment the iteration counter $t = 1$. This completes the first iteration of PSO. The change of swarm position after the first iteration is displayed in Figure 10.6. Since $t = 1 < 100$, we proceed to step 2 (second iteration).

Step 2 (Second Iteration): Fitness Evaluation

In the new position $X^{(1)}$ of the swarm, the values of objective function $f(X^{(1)})$ are evaluated, which are shown in Table 10.4.

Step 3 (Second Iteration): Updating pbest

The fitness value of each particle in its new position is compared with the fitness value of its previous best position. In this example, we observed that

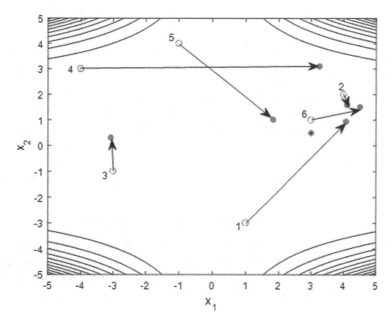

FIGURE 10.6
Change of swarm position after first iteration.

TABLE 10.4

Fitness Values, *pbest*, and *gbest* of Swarm after First Iteration

Particle Number	$X^{(1)}$	$f(X^{(1)})$	$pbest^{(1)}$	$gbest^{(1)}$
1	[4.0778 0.93213]	7.8605	[4.0778 0.93213]	[4.0778 0.93213]
2	[4.1097 1.5814]	304.13	[4.1097 1.5814]	[4.0778 0.93213]
3	[−3.0672 0.30055]	70.153	[−3.0672 0.30055]	[4.0778 0.93213]
4	[3.2672 3.0991]	10,325	[3.2672 3.0991]	[4.0778 0.93213]
5	[1.8293 1.0025]	14.331	[1.8293 1.0025]	[4.0778 0.93213]
6	[4.5 1.5]	253.3	[3 1]	[4.0778 0.93213]

except for particle number 6, all other particles in the swarm show better fitness values in their new positions than their previous best positions. Hence, for particles 1–5, we assign their new positions as *pbest*, and in case of particle number 6, its previous position is assigned as *pbest*. The updated values of *pbest* are put into Table 10.4.

Step 4 (Second Iteration): Updating gbest

It is manifested from Table 10.4 that particle number 1 at its new position [4.0778 0.93213] gives the minimum value of the fitness function among the swarm. Hence, the updated *gbest*$^{(1)}$ is assigned to [4.0778 0.93213].

TABLE 10.5

Updated Velocities of Swarm in Second Iteration

Particle Number	$V^{(1)}$	r_1	r_2	$V^{(2)}$
1	[3.0778 3.9321]	[0.013283 0.19666]	[0.89719 0.093371]	[1.5389 1.9661]
2	[0.10965 −0.41856]	[0.30737 0.10167]	[0.45606 0.99539]	[0.025729 −1.5018]
3	[−0.06725 1.3006]	[0.33209 0.062045]	[0.29735 0.29824]	[4.2155 1.027]
4	[7.2672 0.099128]	[0.046351 0.76143]	[0.50543 0.63107]	[4.453 −2.6855]
5	[2.8293 −2.9975]	[0.089892 0.77724]	[0.080862 0.90513]	[1.7783 −1.6261]
6	[1.5 0.5]	[0.53377 0.82581]	[0.10915 0.3381]	[−0.94349 −0.9598]

Step 5 (Second Iteration): Updating Velocity of Each Particle

In order to update the velocity of each particle, it is necessary to generate random numbers r_1 and r_2. Table 10.5 demonstrates the randomly generated values of r_1 and r_2 for all particles. By applying Equation 10.1, the updated velocity of the first particle can be worked out as follows:

$$V_1^{(2)} = wV_1^{(1)} + C_1 r_1 (pbest_1^{(1)} - X_1^{(1)}) + C_2 r_2 (gbest^{(1)} - X_1^{(1)})$$

$$= \begin{bmatrix} 0.5 \times 3.0778 \\ 0.5 \times 3.9321 \end{bmatrix}^T + \begin{bmatrix} 2 \times 0.013283 \times (4.0778 - 4.0778) \\ 2 \times 0.19666 \times (0.93213 - 0.93213) \end{bmatrix}^T$$

$$+ \begin{bmatrix} 2 \times 0.89719 \times (4.0778 - 4.0778) \\ 2 \times 0.093371 \times (0.93213 - 0.93213) \end{bmatrix}^T$$

$$= \begin{bmatrix} 1.5389 \\ 1.9661 \end{bmatrix}^T + \begin{bmatrix} 2 \times 0.013283 \times 0 \\ 2 \times 0.19666 \times 0 \end{bmatrix}^T + \begin{bmatrix} 2 \times 0.89719 \times 0 \\ 2 \times 0.093371 \times 0 \end{bmatrix}^T$$

$$= \begin{bmatrix} 1.5389 \\ 1.9661 \end{bmatrix}^T$$

$$= [1.5389 \quad 1.9661]$$

The updated velocities of the other particles in the swarm can also be worked out in a similar way. Table 10.5 shows the updated velocities of the whole swarm.

Step 6 (Second Iteration): Updating Position of Each Particle

Each particle in the swarm updates its new position. By applying Equation 10.2, the updated position of the first particle can be calculated as follows:

$$X_1^{(2)} = X_1^{(1)} + V_1^{(2)}$$

$$= [4.0778 \quad 0.93213] + [1.5389 \quad 1.9661]$$

$$= [5.6167 \quad 2.8982]$$

TABLE 10.6

New Position of Swarm after Second Iteration

Particle Number	$X^{(1)}$	$V^{(2)}$	$X^{(2)}$
1	[4.0778 0.93213]	[1.5389 1.9661]	[5 2.8982]
2	[4.1097 1.5814]	[0.025729 −1.5018]	[4.1354 0.079567]
3	[−3.0672 0.30055]	[4.2155 1.027]	[1.1483 1.3276]
4	[3.2672 3.0991]	[4.453 −2.6855]	[5 0.41364]
5	[1.8293 1.0025]	[1.7783 −1.6261]	[3.6076 −0.62364]
6	[4.5 1.5]	[−0.94349 −0.9598]	[3.5565 0.5402]

In this example, the problem space is bounded by $-5 \leq x_1$, $x_2 \leq 5$, and any particle cannot fly beyond this boundary; hence, the new position of the first particle is adjusted to [5   2.8982]. Similarly, the updated positions of other particles in the swarm can be worked out. Table 10.6 shows the new positions of the swarm. We set the iteration counter $t = 2$. Thus, the second iteration of PSO is completed. The change of swarm position after the second iteration is depicted in Figure 10.7.

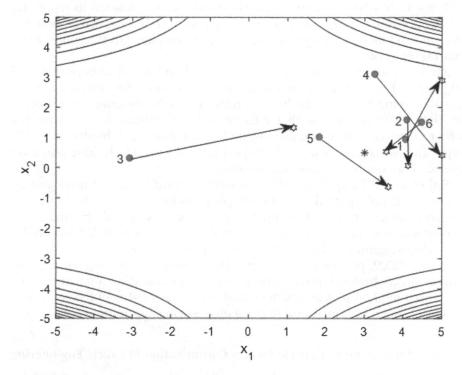

FIGURE 10.7
Change of swarm position after second iteration.

Since $t = 2 < 100$, we proceed to step 2 (third iteration). Steps 2–6 are repeated until the maximum iteration is reached.

The MATLAB® coding for solving the optimization problem given in Equation 10.4 using PSO is presented in Section 10.5.

10.4 Application of Particle Swarm Optimization in Textiles

Although PSO is a popular stochastic search method in other engineering fields, it is very new in the textile domain. Only a few works have been reported on the application of PSO in the domain of textile engineering.

Wang et al. (2010) proposed a new approach of using a PSO algorithm to cluster fabric comfort. The analysis of the clustering results and the comparison of fuzzy cluster results and PSO-based cluster results show that the latter method is feasible and has great practical value.

Liu et al. (2008) applied PSO in backpropagation neural network training for the classification of fabric defect. The method of orthogonal wavelet transform was used to decompose a monolayer from the fabric image. The subimages of horizontal and vertical directions are extracted to represent, respectively, the textures of fabric in warp and weft. In comparison with the neural network alone, they observed favorable results with the hybrid PSO-neural network.

Zhang et al. (2017) proposed a local search enhanced bi-objective PSO algorithm for scheduling textile dyeing processes. They formulated the textile dyeing process scheduling problem as a bi-objective optimization model, in which one objective is connected with tardiness cost, while the other objective reflects the level of pollutant emission. A bi-objective PSO algorithm enhanced by a problem-specific local search technique was used to seek high-quality nondominated solutions.

Raj et al. (2007) applied PSO as a motor controller to optimize energy consumption of a partial loaded three-phase induction motor used in a ring spinning machine. They investigated the economics of a 100 HP induction motor with and without a PSO controller. The study shows the PSO controller introduces significant differences in terms of energy cost.

Zhang (2009) proposed an optimization design of damping for a three-dimensional braided textile composite using the PSO algorithm. The proposed method provides better damping characteristics of the composite material, and it may be useful in other engineering applications.

10.4.1 Application of Particle Swarm Optimization in Fabric Engineering

The term *fabric engineering* may be referred to as the manufacturing of fabrics with desired quality at low costs of production by selecting the appropriate parameters. With appropriate parameters, achieving the requisite quality and keeping the cost of production low simultaneously is possible in any fabric manufacturing process and may be envisaged as the engineering design of fabrics, which is the end goal for every fabric manufacturer. Optimizing fabric parameters like yarn count, crimp, and thread spacing, both in warp and weft directions, holds the key to achieving this target. The intricate relationship of fabric parameters with its physical and mechanical properties makes it too complex to solve this optimization problem by means of a traditional technique. Das et al. (2013, 2014) used the PSO method to search for the best combinations of parameters to obtain fabrics with desired quality with low manufacturing cost. It is necessary to have an appropriate areal density or GSM (g/m²) of fabric for good quality as well as economical production. Fabric with very low GSM is too sparse to qualify for the desirable quality, whereas higher GSM warrants higher manufacturing cost entailing more yarn consumption. Nevertheless, sometimes fabrics manufactured by fine yarns are costlier than the heavy fabrics due to the higher price of fine yarns. Therefore, it is necessary to maintain a particular range of yarn count while optimizing fabric cost by minimizing GSM. Three different ranges of counts for warp and weft yarns—fine (10–24 tex), medium (30–50 tex), and coarse (60–100 tex)—made from 100% cotton fibers were considered for the production of light, medium, and heavyweight fabrics, respectively. The optimization problem was formulated to minimize the GSM (W) of these fabrics such that different physical and mechanical parameters both in warp and weft directions, viz., cover (K), tensile modulus (E), shear rigidity (G), and flexural modulus (B), are retained within the desired levels and the interyarn forces in the fabric are balanced in its relaxed state. The expressions of W, K, E, G, and B are given as follows (Leaf and Kandil, 1980; Leaf and Sheta, 1984; Leaf et al., 1993):

$$W = \frac{T_1(1+c_1)}{p_1} + \frac{T_2(1+c_2)}{p_2} \tag{10.5}$$

$$K = \frac{d_1}{p_1} + \frac{d_2}{p_2} - \frac{d_1 d_2}{p_1 p_2} \tag{10.6}$$

$$E_1 = \frac{12 p_2 \beta_1}{p_1 (l_1 - D\Phi_1)^3 \sin^2 \theta_1} \times \left[1 + \frac{\beta_2 (l_1 - D\Phi_1)^3 \cos^2 \theta_1}{\beta_1 (l_2 - D\Phi_2)^3 \cos^2 \theta_2} \right] \tag{10.7}$$

$$E_2 = \frac{12 p_1 \beta_2}{p_2 (l_2 - D\Phi_2)^3 \sin^2 \theta_2} \times \left[1 + \frac{\beta_1 (l_2 - D\Phi_2)^3 \cos^2 \theta_2}{\beta_2 (l_1 - D\Phi_1)^3 \cos^2 \theta_1} \right] \tag{10.8}$$

$$G = 12 \left\{ \frac{p_1 (l_1 - D\Phi_1)^3}{\beta_1 p_2} + \frac{p_2 (l_2 - D\Phi_2)^3}{\beta_2 p_1} \right\}^{-1} \tag{10.9}$$

$$B_1 = \frac{\beta_1 p_2}{p_1(l_1 - D\Phi_1)} \tag{10.10}$$

$$B_2 = \frac{\beta_2 p_1}{p_2(l_2 - D\Phi_2)} \tag{10.11}$$

where subscripts 1 and 2 refer to parameters in warp and weft directions, respectively; d is the yarn diameter (mm); D is the sum of warp and weft diameters; β is yarn bending rigidity (mN mm²); θ is the weave angle (radians); Φ is the angle of contact (radians); l is the modular length of the yarn in fabric (mm); p is thread spacing (mm); c is yarn crimp in the fabric (fraction); and T is yarn count (tex). The following empirical relationship between θ and c provides a reasonable agreement between theoretical and experimental values (Leaf, 2003):

$$\theta = 1.88c^{1/2} \tag{10.12}$$

Leaf (2003) established the relationship between angle of contact (Φ) and weave angle (θ) for the rigid thread model as follows:

$$\Phi = \sin^{-1}\left(2\sin\theta - \frac{p}{D}\right) \tag{10.13}$$

The yarn diameter (d in mm) is related to its tex (T) by the following expression (Peirce, 1937):

$$d = 0.0357\sqrt{v_y T} \tag{10.14}$$

where v_y is yarn specific volume (cc/g). For cotton yarn, the value of v_y is 1.1 cc/g.

The expression of yarn bending rigidity is given by (Postle et al., 1988)

$$\beta = \beta_f t_f^2 n \tag{10.15}$$

where β_f is specific fiber rigidity (mN mm²/tex²); t_f is fiber (tex); and n is the number of fibers in the yarn cross section. For cotton fiber, the value of β_f is 0.53 mN mm²/tex² (Morton and Hearle, 1993).

At the relaxed state of a fabric, the internal forces acting vertically on the warp and weft threads due to the resistance of bending are balanced if the following equation is satisfied (Peirce, 1937):

$$\frac{\beta_1 \sin\theta_1}{p_2^2} = \frac{\beta_2 \sin\theta_2}{p_1^2} \tag{10.16}$$

Equation 10.16 has predominant influence on the balance of crimp and is therefore known as the crimp balance equation.

The following optimization problem was developed to select the fabric governing parameters such as yarn counts (T_1 and T_2), crimps (c_1 and c_2), and threads spacing (p_1 and p_2) for the production of light, medium, and heavyweight fabrics so as to manufacture them with minimal cost as well as requisite quality:

$$\text{Minimize } W \tag{10.17}$$

subject to the inequality constraints:

$$K^U \geq K \geq K^L$$
$$E_1^U \geq E_1 \geq E_1^L$$
$$E_2^U \geq E_2 \geq E_2^L$$
$$G^U \geq G \geq G^L$$
$$B_1^U \geq B_1 \geq B_1^L$$
$$B_2^U \geq B_2 \geq B_2^L$$

and equality constraint:

$$\frac{\beta_1 \sin \theta_1}{p_2^2} = \frac{\beta_2 \sin \theta_2}{p_1^2}$$

where the superscripts L and U refer to the values of lower and upper bounds, respectively. A penalty function method is used to handle the constraints. A bracket penalty operator was used to handle the inequality constraints, whereas the equality constraint was handled by a parabolic penalty operator (Deb, 2005). Accordingly, for minimizing the objective function $f(x)$ subject to equality constraints $h(x) = 0$ and inequality constraints $g(x) \leq 0$, the penalty function was formulated as

$$\underbrace{P(x)}_{\text{Min}} = f(x) + \mu[\{h(x)\}^2 + \{\max(0, g(x)\}^2] \tag{10.18}$$

where μ is the penalty parameter. The optimization problem of Equation 10.17 was solved using the PSO algorithm.

The boundary for governing parameters as well as the physical and mechanical properties of different fabrics have been selected based on the data collected from the industrial survey. The search space for the governing parameters determining the physical and mechanical properties of different fabrics is given in Table 10.7. Table 10.8 shows the lower and upper bounds of inequality constraints for three types of fabrics.

A population of 500 particles was initialized with random positions and velocities, and the penalty function of Equation 10.18 is evaluated for the whole population. The values of *pbest* are updated, and the best of *pbest* is

TABLE 10.7

Search Space for the Fabric Governing Parameters

Fabric Parameters	Lightweight Fabric (Using Fine Yarns)		Medium-Weight Fabric (Using Medium Count Yarn)		Heavyweight Fabric (Using Coarse Yarns)	
	Lower Bound	Upper Bound	Lower Bound	Upper Bound	Lower Bound	Upper Bound
T_1, tex	10	24	30	50	60	100
T_2, tex	10	24	30	50	60	100
c_1, fraction	0.03	0.15	0.03	0.15	0.03	0.15
c_2, fraction	0.03	0.15	0.03	0.15	0.03	0.15
p_1, mm	0.20	0.40	0.45	0.80	0.50	0.85
p_2, mm	0.20	0.40	0.45	0.80	0.50	0.85

Source: Das, S. et al. 2014. *International Journal of Clothing Science and Technology*, 26(1), 48–57.

TABLE 10.8

Boundary of the Constraints for Different Types of Fabrics

Fabric Parameters	Lightweight Fabric (Using Fine Yarns)		Medium-Weight Fabric (Using Medium Count Yarn)		Heavyweight Fabric (Using Coarse Yarns)	
	Lower Bound	Upper Bound	Lower Bound	Upper Bound	Lower Bound	Upper Bound
K	0.6	0.72	0.6	0.72	0.6	0.72
E_1, mN/mm	1,000	2,500	2,000	3,500	3,500	5,500
E_2, mN/mm	1,000	2,500	2,000	3,500	3,500	5,500
G, mN/mm	50	150	100	250	200	350
B_1, mN.mm	3	12	6	15	12	20
B_2, mN.mm	3	12	6	15	12	20

Source: Das, S. et al. 2014. *International Journal of Clothing Science and Technology*, 26(1), 48–57.

then set as *gbest*, and the value of inertial weight is adjusted as per Equation 10.3. The velocity and position are then modified according to Equations 10.1 and 10.2. This completes the first iteration of PSO. The PSO algorithm runs until it reaches the maximum number of iterations, which was set to 1,000. The values of C_1, C_2, w_{max}, and w_{min} were set to 2, 2, 1.2, and 0.2, respectively.

The optimum values of the fabric governing parameters obtained by solving the optimization problem of Equation 10.17 for lightweight, medium weight, and heavyweight fabrics are summarized in Table 10.9. The physical and mechanical properties of fabrics were predicted from the obtained parameters of Table 10.9 using Equations 10.5 and 10.11. Table 10.10 depicts the predicted values of the physical and mechanical properties of different fabrics. It is evident from Table 10.10 that the predicted properties of fabrics are lying within the desirable limits. The optimized fabric parameters given in Table 10.9 also satisfy Equation 10.16. This ensures the balance of interyarn forces in the fabric,

TABLE 10.9

Optimum Values of Governing Parameters

Fabric Type	T_1 (tex)	T_2 (tex)	c_1 (fraction)	c_2 (fraction)	p_1 (mm)	p_2 (mm)
Lightweight	13	13	0.100	0.110	0.400	0.390
Medium weight	32	32	0.109	0.113	0.621	0.618
Heavyweight	72	72	0.117	0.113	0.789	0.792

Source: Das, S. et al. 2014. *International Journal of Clothing Science and Technology*, 26(1), 48–57.

TABLE 10.10

Obtained Fabric Properties

Fabric Type	K	E_1, mN/mm	E_2, mN/mm	G, mN/mm	B_1, mN.mm	B_2, mN.mm	W, g/m²
Lightweight	0.57	1,848	1,718	145.3	3.0	3.0	72.7
Medium weight	0.57	2,381	2,293	200.7	6.0	6.0	114.8
Heavyweight	0.64	3,502	3,640	316.2	12.0	12.0	203.1

Source: Das, S. et al. 2014. *International Journal of Clothing Science and Technology*, 26(1), 48–57.

which in turn affects the balance of warp and weft crimp. The obtained values of fabric GSM with the optimized parameters are 72.7, 114.8, and 203.7 g/m² for lightweight, medium weight, and heavyweight fabrics, respectively.

For experimental validation of the theoretical results, a plain woven fabric was prepared in a sample loom using the optimized parameters obtained by the PSO algorithm for the medium-weight fabric as given in Table 10.9. Manufacturing of the fabric sample with the required specifications was not easy; in particular, controlling crimp during fabric production was a very difficult job. Therefore, the manufacture of several fabric samples was tried out in the weaving laboratory to produce them as close as possible to the required specifications as obtained with PSO. Eventually, fabric with the closest parameters has been considered for validation of the work. Actual crimp of the selected fabric in the relaxed state was 11% and 11.2% in warp and weft, respectively. Dry relaxation of the fabric was carried out at standard testing atmosphere (27°C, 65% r.h.) for 24 hours before the experiment.

Fabric GSM (W) was measured according to ASTM standards (ASTM Standard, 2002). Warp and weft thread densities were measured using a pick glass. The warp and weft counts were determined using a Beesely yarn balance. An average of 10 readings was taken to estimate the fabric cover (K) from the experimental results of thread densities and counts.

The tensile moduli (E_1 and E_2 in mN/mm) of fabric were obtained by extending the fabric sample of size 25 cm × 5 cm on an Instron tensile tester with a cross-head speed of 2 cm/min. The tangent at the origin to the resulting load-elongation curve was taken as a measure of the fabric initial modulus. Five separate readings were taken in both warp and weft directions to determine the average values of E_1 and E_2.

The flexural moduli (B_1 and B_2 in mN mm) of the fabric were measured using the following equation:

$$B = W \times C^3 \times 9.807 \times 10^{-3}$$
(10.19)

where C is the bending length (cm). The Shirley stiffness tester was used to determine the bending length of a fabric sample 25 mm wide and 200 mm long in both warp and weft directions. Three samples in each direction were cut so that no two warp specimens contained the same warp threads, and no two weft specimens contained the same weft threads. Four readings were taken from each specimen, one face up and one face down on the first end, and then the same for the second end. The average values of bending length in warp and weft directions were measured to be 1.83 and 1.81 cm, respectively.

Shear rigidity (G in mN/mm) of fabric was estimated using the following equation (Leaf and Sheta, 1984):

$$\frac{1}{G} = \frac{4}{E_{45}} - \frac{(1-\sigma_2)}{E_1} - \frac{(1-\sigma_1)}{E_2}$$
(10.20)

where E_{45} is the tensile modulus of the fabric in a direction making an angle of 45° with the warp threads, and σ_1 and σ_2 are the fabric Poisson's ratios in warp and weft directions, respectively. Poisson's ratios were measured by photographing the fabric at regular intervals during the tensile testing in the Instron tensile tester. The changes in dimensions of a rectangle marked on the fabric specimen were measured from the photographs, which led to the estimation of fabric Poisson's ratios. The measured values of E_{45}, σ_1, and σ_2 were 698 mN/mm, 0.59, and 0.34, respectively.

Table 10.11 shows the calculated and measured values of physical and mechanical properties of fabric produced from the optimized parameters. Table 10.11 shows significant agreement between the calculated and measured values of fabric properties.

TABLE 10.11

Comparison of Experimental Values of Fabric Properties with Calculated Values

Fabric Properties	Calculated Values	Measured Values	Error (%)
K	0.57	0.58	1.72
E_1, mN/mm	2,381	2,305	3.29
E_2, mN/mm	2,293	2,188	4.79
G, mN/mm	200.7	190.2	5.52
B_1, mN.mm	6.0	6.5	7.69
B_2, mN.mm	6.0	6.3	4.76
W, g/m²	114.8	108	6.29

Source: Das, S. et al. 2014. *International Journal of Clothing Science and Technology*, 26(1), 48–57.

Thus, PSO is found to be a useful nontraditional optimization technique for the engineering design of fabrics, since it can easily be implemented and is computationally inexpensive as fewer parameters are involved. Using this system, a fabric manufacturer can efficiently determine the required parameters, such as count, thread spacing, and crimp, for manufacturing the fabrics with desired properties at an optimum cost. Thus, fabric manufacturers can benefit from the quick response capability provided by this system and can then enhance their own competitive abilities.

10.5 MATLAB® Coding

MATLAB® Coding for PSO

```
% Program of Particle Swarm Optimization
clc
clear all
close all
format short g
 % Initialize the parameters
m = 2; % number of variables
n = 1000; % swarm size
w = 0.5;  % inertia weight
c1 = 2; % learning factor 1
c2 = 2; % learning factor 2
ub = [5 5]; % upper bound of variables
lb = [-5 -5]; % lower bound of variables
itr = 100;% number of iterations
 % Initialize the swarm position
for i = 1: n
    for j = 1: m
        x0(i, j) = (lb(j) + rand()*(ub(j) - lb(j)));
    end
end
v = x0; % initial swarm velocity
% Fitness of initial swarms
fx0 = (1.5 - x0(:,1) + x0(:,1).*x0(:,2)).^2 + (2.25 - x0(:,1)
+ x0(:,1).*x0(:,2).^2).^2 ...
        + (2.625 - x0(:,1) + x0(:,1).*x0(:,2).^3).^2;
pbest = x0; % initial pbest
[fmin0,index0] = min(fx0);
gbest = x0(index0,:); % initial gbest
x = x0; % initial swarm population

% Particle swarm optimization starts.....
% Velocity update of swarm
```

```
for j = 1: itr
      for i = 1: n
      for j = 1: m
            v(i,j) = w*v(i, j) + c1*rand()*(pbest(i,j)
- x(i,j))...
                        + c2*rand()*(gbest(1,j) - x(i,j));
        end
    end
     % Swarm position update
    for i = 1: n
        for j = 1: m
            x(i,j) = x(i,j) + v(i,j);
        end
    end
     % Handling boundary violations
    for i = 1: n
        for j = 1: m
            if x(i,j) < lb(j)
                x(i,j) = lb(j);
            elseif x(i,j) > ub(j)
                x(i,j) = ub(j);
            end
        end
    end
     % Evaluate fitness of new swarms
    fx = (1.5-x(:,1) + x(:,1).*x(:,2)).^2 + (2.25 - x(:,1) +
x(:,1).*x(:,2).^2).^2 ...
        + (2.625 - x(:,1) + x(:,1).*x(:,2).^3).^2;
     % Updating pbest and fitness
    for i =1: n
        if fx(i, 1) < fx0(i, 1)
            pbest(i, :) = x(i, :);
            fx0(i, 1) = fx(i, 1);
        end
    end
    [fmin, index] = min(fx0); % finding out the best particle
     % Updating gbest and best fitness
    if fmin < fmin0
       gbest = pbest(index,:);
       fmin0 = fmin;
    end
end
% Optimum values of variables
gbest
% Optimum function value
fn_val = (1.5 - gbest(:, 1) + gbest(:, 1).*gbest(:, 2)).^2 ...
        + (2.25 - gbest(:, 1) + gbest(:, 1).*gbest(:,
2).^2).^2 ...
        + (2.625 - gbest(:, 1) + gbest(:, 1).*gbest(:,
2).^3).^2
```

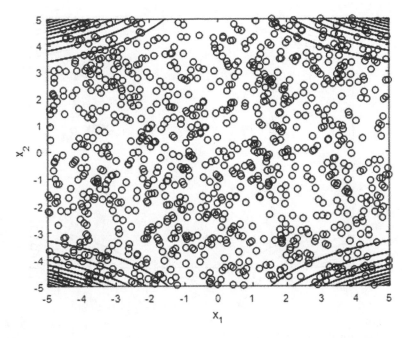

FIGURE 10.8
Spread of initial swarm of size 1,000 on the contour plot.

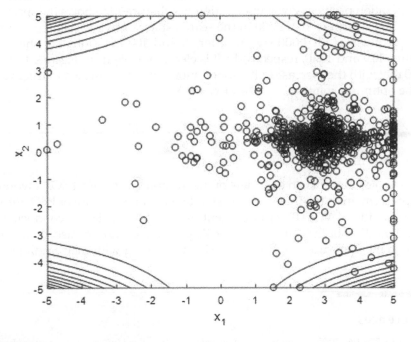

FIGURE 10.9
Positions of swarm after 10th iteration.

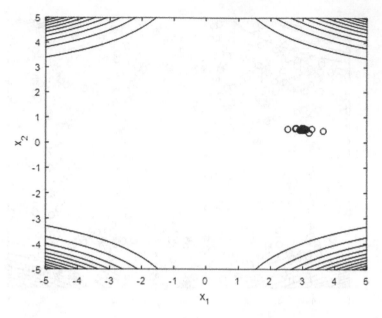

FIGURE 10.10
Positions of swarm after 100th iteration.

The MATLAB® code of PSO was executed with 1,000 swarm to solve the optimization problem given in Equation 10.4. Figure 10.8 shows the spread of the initial swarm of size 1,000 in the search space of two variables x_1 and x_2. The positions of all 1,000 swarm after 10 and 100 iterations are shown in Figures 10.9 and 10.10, respectively. It is clearly visible from Figures 10.9 and 10.10 that with the increase of iterations, more and more swarm will converge to the optimum solution $x_1 = 3$ and $x_2 = 0.5$.

10.6 Summary

This chapter presents an overview of the different steps of a PSO. A worked-out problem analytically shows the step-by-step architecture of PSO with its coding in the MATLAB® environment. A literature review is conducted on the application of PSO in various textile problems. Authors also shed a new light on the application of PSO in fabric engineering with their findings.

References

ASTM D3776-96. 2002. *Standard Test Methods for Mass Per Unit Area (Weight) of Fabric*, ASTM International, West Conshohocken, PA.

Das, S., Ghosh, A. and Banerjee, D. 2013. Engineering design of woven fabrics using non-traditional optimization methods: A comparative study. *Fibers and Polymers*, 14(9), 1562–1567.

Das, S., Ghosh, A. and Banerjee, D. 2014. Designing of engineered fabrics using particle swarm optimization. *International Journal of Clothing Science and Technology*, 26(1), 48–57.

Deb, K. 2005. *Optimization for Engineering Design Algorithms and Examples*, Prentice Hall, New Delhi, India.

Kennedy, J. and Eberhart, R. C. 1995. Particle swarm optimization. *Proceedings of IEEE International Conference on Neural Networks*, Perth, Australia, 1942–1948.

Leaf, G. A. V. 2003. The mechanics of plain woven fabrics. *Proceedings of International Textile Design and Engineering Conference INTEDEC*, Edinburgh, United Kingdom.

Leaf, G. A. V., Chen, Y. and Chen, X. 1993. The initial bending behaviour of plain woven fabrics. *Journal of the Textile Institute*, 84(3), 419–428.

Leaf, G. A. V. and Kandil, K. H. 1980. The initial load-extension behaviour of plain woven fabric. *Journal of the Textile Institute*, 71(1), 1–7.

Leaf, G. A. V. and Sheta, A. M. F. 1984. The initial shear modulus of plain-woven fabrics. *Journal of the Textile Institute*, 75(3), 157–163.

Liu, S., Liu, J. and Zhang, L. 2008. Classification of fabric defect based on PSO-BP neural network. *Proceedings of the Second International Conference on Genetic and Evolutionary Computing*, Hubei, China, 137–140.

Morton, W. E. and Hearle, J. W. S. 1993. *Physical Properties of Textile Fibers*, 3rd ed., The Textile Institute, Manchester.

Peirce, F. T. 1937. The geometry of cloth structure. *Journal of the Textile Institute*, 28(3), T45–T112.

Postle, R., Caranby, G. A. and Jong, S. 1988. *The Mechanics of Wool Structures*, Ellis Horwood, Chichester, England.

Raj, T. C., Agarwal, P. and Srivastava, S. P. 2007. Particle swarm optimized induction motor for a textile mill load diagram. *Proceedings of the International Conference on Information and Communication Technology in Electrical Sciences*, Tamil Nadu, India.

Shi, Y. and Eberhart, R. C. 1998a. A modified particle swarm optimizer. *Proceedings of IEEE International Conference on Computational Intelligence*, Anchorage, Alaska, May 4–9, 69–73.

Shi, Y. and Eberhart, R. C. 1998b. Parameter selection in particle swarm adaptation. *Proceedings of the Seventh International Conference on Evolutionary Programming*, San Diego, CA, 591–600.

Umapathy, P., Venkataseshaiah, C. and Senthil, M. 2010. Particle swarm optimization with various inertia weight variants for optimal power flow solution. *Discrete Dynamics in Nature and Society*, 10, 1–15.

Wang, D., Zhang, L. and Zeng, P. 2010. Clustering research of fabric deformation comfort using bi-swarm PSO algorithm. *Proceedings of the Eighth World Congress on Intelligent Control and Automation*, Jinan, China.

Zhang, K. 2009. Optimization of the damping of the rectangular 3-D braided composite based on PSO algorithm. *Proceedings of the Fourth International Symposium*, Huangshi, China, 454–465.

Zhang, R., Chang, P. C., Song, S. and Wu, C. 2017. Local search enhanced multi-objective PSO algorithm for scheduling textile production processes with environmental considerations. *Applied Soft Computing*, 61, 447–467.

11

Simulated Annealing

11.1 Introduction

Simulated annealing was first put forward by Metropolis et al. (1953) and successfully applied to the optimization problems by Kirkpatrick et al. (1983). It emulates the cooling process of molten metals through annealing. It is a random search technique that exploits an analogy between the way in which a metal cools and freezes into a minimum energy crystalline structure. It begins with a starting temperature and by carefully controlling the rate of cooling it may lead to a global optimum. Unlike the genetic algorithm and particle swarm optimization which are population-based search methods, simulated annealing is a point-by-point search method.

Simulated annealing starts with a randomly selected current point in the search space, and the initial temperature T is kept at a high value. A new point is generated at random in the neighborhood of the current point. Then the difference in the function values ΔE at these two points is measured, and the Metropolis algorithm is used to accept or reject the point. If the new point is accepted, it is used as a starting point for the next step. However, if the new point is rejected, the original point is retained as a starting point of the next step. This is the end of the first iteration of simulated annealing. In the next iteration, again a new point is randomly generated in the vicinity of the current point, and the Metropolis algorithm is used to accept or reject the new point. In general, before reducing the temperature, a number of iterations are performed at a particular temperature. The temperature is lowered according to the cooling schedule, and the rate of cooling is done at a slower rate. The algorithm is stopped when the temperature becomes just lower than the termination temperature.

11.2 Simulated Annealing

Annealing means the tempering of a metal by heating above its melting point and then cooling it very slowly until it solidifies into a perfect crystalline structure. The simulation of the annealing process is termed as *simulated annealing* (Kirkpatrick et al., 1983). At high temperature, the atoms in a molten metal have greater mobility, and they can move freely with respect to each other, but as the temperature is lowered, the movement of the atoms gets restricted. A fundamental question in statistical mechanics concerns the ultimate state of the material in the limit of low temperature: do the atoms remain fluid or solidify, and if they solidify, do they form a perfect crystalline solid, an imperfect crystalline solid, or a glass? The rate at which the temperature is reduced plays a key role in deciding the ultimate state of the material. The growth of a perfect crystal from a molten metal is achieved by careful annealing in which the temperature is lowered very slowly and a long time is spent at temperatures in the vicinity of the freezing point. Because of very slow cooling, the atoms get ordered gradually, and finally they form a crystal having the minimum energy level. Therefore, in order to achieve the absolute minimum energy level, the temperature needs to be reduced at a very slow rate. On the contrary, if the temperature of a molten metal is reduced at a faster rate, the resulting crystals will have many defects or may become a glass with no crystalline order. A defective crystal or a glass has a higher energy level. Figure 11.1 illustrates the effect of cooling rate on the ultimate state of a molten metal. A defective crystalline structure corresponds to local minimum configurations, whereas a perfect crystalline structure corresponds to a global minimum energy configuration.

Simulated annealing is analogous to an optimization procedure such that the energy of a state is the value of an objective function in a minimization

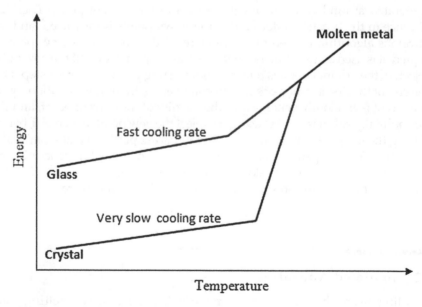

FIGURE 11.1
Effect of cooling rate.

problem, and the cooling phenomenon is simulated by controlling a temperature-like parameter introduced with the concept of a Boltzmann probability distribution (Deb, 2005). In statistical mechanics, the Boltzmann probability distribution gives the probability p_i of a physical system being in state i with energy E_i at temperature T by the following expression:

$$p_i = \frac{e^{-E_i/kT}}{\sum_{j=1}^{m} e^{-E_j/kT}} \tag{11.1}$$

where k is the Boltzmann constant, and m is the number of states accessible to the system.

This shows that states with lower energy will always have a higher probability of being occupied than the states with higher energy. According to the Boltzmann distribution, the probability of a state change p can be determined from the energy difference of the two states ΔE and temperature T as follows:

$$p = e^{-\Delta E/kT} \tag{11.2}$$

Figure 11.2 displays the plot of probability of state change with temperature, from which it is quite evident that at a high temperature, there is uniform

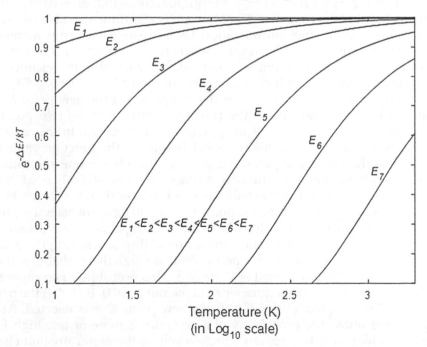

FIGURE 11.2
Probability of state change depending on the energy difference and temperature.

preference for all of the states, irrespective of energy; however, at a low temperature, there is a small probability of being at a high energy state. Thus, when the temperature approaches absolute zero, only the states with minimum energy have nonzero probability of occurrence. In simulated annealing, the search process follows the Boltzmann probability distribution where Boltzmann constant k is omitted and convergence of the algorithm is obtained by controlling the temperature T. The implementation of the Boltzmann probability distribution can be performed by using the Metropolis algorithm, which is a Monte Carlo simulation to generate sample states of a simulated thermodynamic system.

Metropolis et al. (1953) introduced a simple algorithm that can provide an efficient simulation of a collection of atoms in equilibrium at a given temperature. According to this algorithm, in each step, an atom is given a small displacement at random, and the resulting change in energy of the system ΔE is computed. If $\Delta E \leq 0$, the displacement is accepted, and the configuration with the displaced atom is used as the starting point of the next step. Otherwise, it is treated probabilistically using the Boltzmann distribution, that is, if $\Delta E > 0$, the probability that the configuration is accepted is $P = e^{-\Delta E/T}$. The random part of the algorithm is implemented by selecting a uniformly distributed random number P' in the range $(0, 1)$. The selected number P' is compared with P. If $P' \leq e^{-\Delta E/T}$, the new configuration is retained, otherwise if $P' > e^{-\Delta E/T}$, the original configuration is used as the starting point of the next step. The Metropolis algorithm can also be used in the context of function minimization. Let us assume that at any moment the current point is $X^{(i)}$, which gives a function value $E^{(i)} = f(X^{(i)})$. Let us also assume that a new point $X^{(i+1)}$ is randomly created at the vicinity of the current point $X^{(i)}$. The function value of the point $X^{(i+1)}$ is $E^{(i+1)} = f(X^{(i+1)})$. The Metropolis algorithm says that the acceptance of the new point $X^{(i+1)}$ depends on the difference in the function values at these two points: $\Delta E = f(X^{(i+1)}) - f(X^{(i)})$. If $\Delta E \leq 0$, new point $X^{(i+1)}$ is accepted. In the context of function minimization, this is logical, because if the function value at $X^{(i+1)}$ is less than that at $X^{(i)}$, then the point $X^{(i+1)}$ is a better one and thus it is accepted. However, the situation becomes probabilistic when $\Delta E > 0$, which means that the function value at $X^{(i+1)}$ is higher than that at $X^{(i)}$. Had it been a traditional algorithm, the point $X^{(i+1)}$ would have not been accepted in this particular situation. But what is interesting about the Metropolis algorithm is that there exists some finite probability of selecting the point $X^{(i+1)}$, even if it is inferior to the point $X^{(i)}$. If $\Delta E > 0$, the probability that the new point $X^{(i+1)}$ is accepted is $P = e^{-\Delta E/T}$. This probability is compared with a randomly generated number P' in the range $(0, 1)$. If $P' \leq P$, the new point $X^{(i+1)}$ is accepted; else if $P' > P$, the new point $X^{(i+1)}$ is rejected. At a high temperature, the probability of acceptance is more or less high for the points with largely disparate function values; therefore, any point has a fairly good chance of acceptance. On the contrary, at a low temperature, the probability of accepting a random point is becoming small; therefore, the points with only small deviation in function value are accepted. With the intention of simulating the thermal equilibrium at every temperature, usually a number of iteration n is performed at a particular temperature. The temperature is then lowered step by step at a slower rate. The algorithm is terminated when temperature becomes just less than the minimum temperature.

11.2.1 Flowchart of Simulated Annealing

Step 1: Initialization

Choose an initial point X. Set initial temperature T to a sufficiently high value, cooling rate C_T, and minimum temperature T_{min}. Also set $i = 0$. Estimate the function value for the current point $E(X)$.

Step 2: Generation of a New Point in the Neighborhood of Current Point

Create a new point $X^{(i+1)}$ randomly at the vicinity of the current point X. Estimate the function value for the new point $E(X^{(i+1)})$.

Step 3: Checking the Acceptance of the New Point

If $\Delta E = E(X^{(i+1)}) - E(X) < 0$, the new point is accepted, set $X = X^{(i+1)}$, else create a random number P' in between the range (0–1). If $P' \leq \exp(-\Delta E/T)$, the new point is accepted, set $X = X^{(i+1)}$, else the new point is rejected.
 Set $i = i + 1$, and go to the next step.

Step 4: Lowering the Temperature as per the Cooling Schedule

Lower the temperature T according to the cooling schedule. If $T < T_{min}$, terminate, or else go to step 2.
 The flowchart of the simulated annealing algorithm is depicted in Figure 11.3.

11.3 Step-by-Step Working Principle of Simulated Annealing

Let us consider the following optimization problem to explain the working principle of simulated annealing:

$$\begin{aligned} \text{Minimize} \quad & f(x_1, x_2) = (x_1 + 2x_2 - 7)^2 + (2x_1 + x_2 - 5)^2 \\ \text{Subject to} \quad & -10 \leq x_1, x_2 \leq 10 \end{aligned} \quad (11.3)$$

where x_1 and x_2 are the real variables.

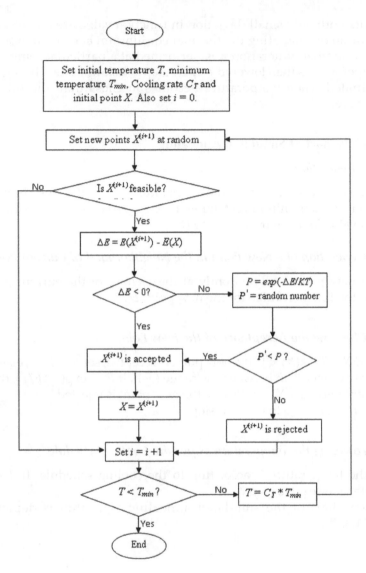

FIGURE 11.3
Flowchart of the simulated annealing algorithm.

Surface and contour plots of the objective function $f(x_1, x_2)$ are displayed in Figures 11.4 and 11.5, respectively. It can be shown that the minimum value of the function $f(x_1, x_2)$ is 0, which is obtained at $x_1 = 1$ and $x_2 = 3$. In both Figures 11.4 and 11.5, the optimum solution is marked by an asterisk (*).

Various steps of simulated annealing up to 10 iterations are discussed as follows for solving the optimization problem of Equation 11.3.

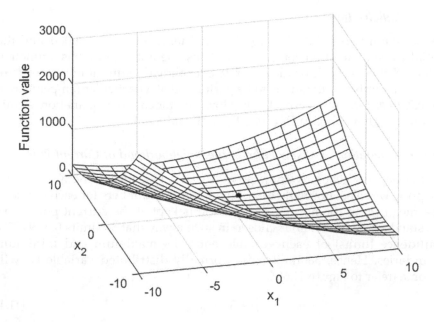

FIGURE 11.4
Surface plot of function $f(x_1, x_2) = (x_1 + 2x_2 - 7)^2 + (2x_1 + x_2 - 5)^2$.

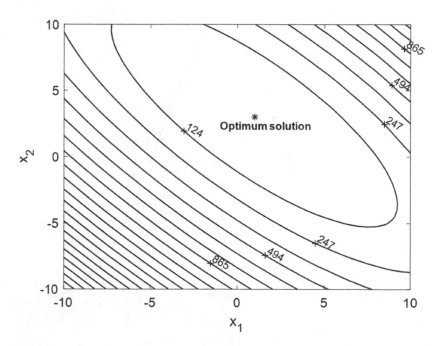

FIGURE 11.5
Contour plot of the function $f(x_1, x_2) = (x_1 + 2x_2 - 7)^2 + (2x_1 + x_2 - 5)^2$.

Step 1: Initialization

We set the initial point $X^{(0)} = (0, 0)$, initial temperature $T = 1,000$, and the initial value of iteration counter $i = 0$. Thus, the initial point has a function value $f(X^{(0)}) = 74$. Typical values of rate of cooling C_T, minimum temperature T_{min}, and number of iterations to be performed at a particular temperature n are 0.9, 10^{-6}, and 1,000, respectively. However, for ease of explanation, in this example we have chosen $C_T = 0.5$, $T_{min} = 1$, and $n = 1$.

Step 2: Generation of a New Point in the Neighborhood of Current Point

In this step, we generate a new point $X^{(1)}$ in the vicinity of the $X^{(0)}$. For this purpose, we select normally distributed random numbers for each variable. The mean (μ) of the normal distribution is kept at the current point, and the standard deviation (σ) is chosen in such a way that 3σ limits (i.e., 99.73% confidence limits) of each variable equal its maximum and minimum boundaries. Hence, 3σ limits of a normally distributed variable (x) will become (refer to Figure 11.6):

$$\mu + 3\sigma = x_{max} \tag{11.4}$$

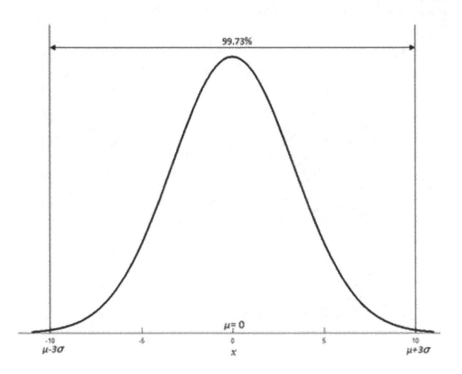

FIGURE 11.6
Normal distribution of variable x with mean $= 0$.

$$\mu - 3\sigma = x_{\min} \tag{11.5}$$

where $x_{\max}$ and $x_{\min}$ are the maximum and minimum boundaries of the variable. By solving Equations 11.4 and 11.5, we get

$$\sigma = \frac{(x_{\max} - x_{\min})}{6} \tag{11.6}$$

In this example, for both the variables (x_1, x_2), the current point $X^{(0)}$, maximum and minimum boundaries are (0, 0), (10, 10), and (−10, −10), respectively. Thus, each variable is assumed to have normal distribution with $\mu = 0$ and $\sigma = 3.3333$. With this distribution, let us suppose that we obtain two random numbers as $x_1 = -7.337$ and $x_2 = 0.53718$. Therefore, the new point is $X^{(1)} = (-7.337, 0.53718)$, and its function value is $f(X^{(1)}) = 542.11$.

Step 3: Checking the Acceptance of the New Point

At this step, we estimate the change in function value which is worked out as $\Delta E = f(X^{(1)}) - f(X^{(0)}) = 542.11 - 74 = 468.11 > 0$. Since $\Delta E > 0$, the acceptance of the new point is decided by the Metropolis algorithm. According to this algorithm, the probability of accepting new point (P) is $e^{-(468.1/1000)} = 0.6262$. We select a random number (P') between 0 and 1. Let us assume that $P' = 0.43091$. Since $P' < P$, the new point $X^{(1)} = (-7.337, 0.53718)$ is turned out to be accepted. We increment the iteration counter $i = 1$. This completes the first iteration of simulated annealing. In actual practice, a large number of iterations up to 1,000 is tried before lowering the temperature.

Step 4: Lowering the Temperature as per the Cooling Schedule

In this step, the temperature is lowered according to the cooling schedule. As we have selected the rate of cooling $= 0.5$, the new temperature (T) is set to $0.5 \times 1{,}000 = 500$. Since $T > T_{\min}$, we go back to step 2.

Step 2: Generation of a New Point in the Neighborhood of Current Point

We randomly generate another point $X^{(2)}$ in the vicinity of the $X^{(1)}$. Using the same procedure as discussed in the preceding section, we generate two random numbers based on the normal distribution of two variables $(x_1$ and $x_2)$ with $\mu = -7.337$ and 0.53718, respectively (refer to Figures 11.7 and 11.8). The value of σ for both variables is kept at 3.3333. Let us assume that we obtain two random numbers as $x_1 = -11.2205$ and $x_2 = 1.8084$. Since the value of variable x_1 is falling beyond the lower boundary, we reject this point. We set the iteration counter $i = 2$ and proceed to step 4. Therefore, after the second iteration, the new point $X^{(2)}$ remains at $X^{(1)}$.

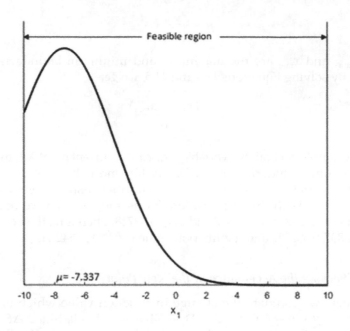

FIGURE 11.7
Normal distribution of variable x_1 with mean $= -7.337$.

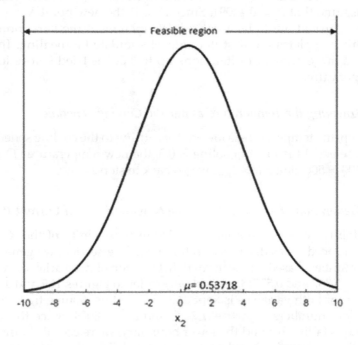

FIGURE 11.8
Normal distribution of variable x_2 with mean $=0.53718$.

Step 4: Lowering the Temperature as per the Cooling Schedule

The temperature (T) is reduced to $0.5 \times 500 = 250$. Since $T > T_{min}$, we go back to step 2.

Step 2: Generation of a New Point in the Neighborhood of Current Point

The new point $X^{(3)}$ is again generated in the neighborhood of the current point $X^{(2)}$ following the same procedure as in the previous iteration. Suppose that the new point is $X^{(3)} = (-8.2649, 0.6315)$ and it has a function value $f(X^{(3)}) = 632.79$.

Step 3: Checking the Acceptance of the New Point

The change in function value is $\Delta E = f(X^{(3)}) - f(X^{(2)}) = 632.79 - 542.11 = 90.68 > 0$. Since $\Delta E > 0$, we generate a random number between the range $(0, 1)$: $P' = 0.25622$. The probability of accepting the new point (P) is $e^{-(90.68/250)} = 0.69578$. Since $P' < P$, the new point $X^{(3)} = (-8.2649, 0.6315)$ will be accepted. We increment the iteration counter $i = 3$ and go to step 4.

Step 4: Lowering the Temperature as per the Cooling Schedule

The temperature (T) is reduced to $0.5 \times 250 = 125$. Since $T > T_{min}$, we go back to step 2.

Step 2: Generation of a New Point in the Neighborhood of Current Point

Suppose that the new point $X^{(4)} = (-10.5867, -1.9691)$ is found in the neighborhood of the current point $X^{(3)}$. This point is rejected; because the value of variable x_1 is falling beyond the lower boundary. We set the iteration counter $i = 4$ and proceed to step 4. Thus, after the fourth iteration, the new point $X^{(4)}$ remains at $X^{(3)}$.

Step 4: Lowering the Temperature as per the Cooling Schedule

The temperature (T) is set to $0.5 \times 125 = 62.5$. Since $T > T_{min}$, we go back to step 2.

Step 2: Generation of a New Point in the Neighborhood of Current Point

The new point $X^{(5)} = (-8.9895, 1.6117)$ is created in the neighborhood of the current point $X^{(4)}$, and it has a function value $f(X^{(5)}) = 619.54$.

Step 3: Checking the Acceptance of the New Point

The change in function value is $\Delta E = f(X^{(5)}) - f(X^{(4)}) = 619.54 - 632.79 < 0$. Since $\Delta E < 0$, point $X^{(5)}$ is accepted. We set iteration counter $i = 5$ and proceed to step 4.

Step 4: Lowering the Temperature as per the Cooling Schedule

The temperature (T) is set to $0.5 \times 62.5 = 31.25$. Since $T > T_{min}$, we proceed to step 2.

Step 2: Generation of a New Point in the Neighborhood of Current Point

The new point $X^{(6)} = (-5.4466, 2.3564)$ is generated in the vicinity of the current point $X^{(5)}$, and it gives a function value $f(X^{(6)}) = 243.05$.

Step 3: Checking the Acceptance of the New Point

The change in function value is $\Delta E = f(X^{(6)}) - f(X^{(5)}) = 243.05 - 619.54 < 0$. Since $\Delta E < 0$, point $X^{(6)}$ will be accepted. Thus, $i = 6$, and we move to step 4.

Step 4: Lowering the Temperature as per the Cooling Schedule

The temperature (T) is set to $0.5 \times 31.25 = 15.625$. Since $T > T_{min}$, we proceed to step 2.

Step 2: Generation of a New Point in the Neighborhood of Current Point

The new point $X^{(7)} = (-5.2255, -1.7712)$ is found in the vicinity of the current point $X^{(6)}$, and it gives a function value $f(X^{(7)}) = 545.22$.

Step 3: Checking the Acceptance of the New Point

The change in function value is $\Delta E = f(X^{(7)}) - f(X^{(6)}) = 545.22 - 243.05 = 302.17 > 0$. Since $\Delta E > 0$, we generate a random number between the range $(0, 1)$: $P' = 0.086539$. The probability of accepting the new point (P) is $e^{-(302.17/15.625)} = 3.9924 \times 10^{-9}$. Since $P' > P$, the new point is rejected. We set the iteration counter $i = 7$ and go to step 4. Therefore, after the seventh iteration, the new point $X^{(7)}$ remains at $X^{(6)}$.

Step 4: Lowering the Temperature as per the Cooling Schedule

The temperature (T) is reduced to $0.5 \times 15.625 = 7.8125$. Since $T > T_{min}$, we proceed to step 2.

Step 2: Generation of a New Point in the Neighborhood of Current Point

The new point $X^{(8)} = (-3.2893, 5.386)$ is found in the vicinity of the current point $X^{(7)}$, which has a function value $f(X^{(8)}) = 38.581$.

Step 3: Checking the Acceptance of the New Point

The change in function value is $\Delta E = f(X^{(8)}) - f(X^{(7)}) = 38.581 - 545.22 < 0$. Since $\Delta E < 0$, point $X^{(8)}$ is accepted. We increment the iteration counter $i = 8$ and move to step 4.

Step 4: Lowering the Temperature as per the Cooling Schedule

The temperature (T) is reduced to $0.5 \times 7.8125 = 3.9063$. Since $T > T_{min}$, we proceed to step 2.

Step 2: Generation of a New Point in the Neighborhood of Current Point

The new point $X^{(9)} = (-1.4429, 2.6077)$ is found in the neighborhood of the current point $X^{(8)}$, which yields a function value $f(X^{(9)}) = 38.275$.

Step 3: Checking the Acceptance of the New Point

The change in function value is $\Delta E = f(X^{(9)}) - f(X^{(8)}) = 38.275 - 38.581 < 0$. Since $\Delta E < 0$, point $X^{(9)}$ is accepted. The iteration counter becomes $i = 9$, and we move to step 4.

Step 4: Lowering the Temperature as per the Cooling Schedule

The temperature (T) is lowered to $0.5 \times 3.9063 = 1.9531$. Since $T > T_{min}$, we go to step 2.

Step 2: Generation of a New Point in the Neighborhood of Current Point

The new point $X^{(10)} = (-5.0640, 0.2023)$ is found in the neighborhood of the current point $X^{(9)}$, and it yields a function value $f(X^{(10)}) = 358.72$.

Step 3: Checking the Acceptance of the New Point

The change in function value is $\Delta E = f(X^{(10)}) - f(X^{(9)}) = 358.720 - 38.275 = 320.44 > 0$. Since $\Delta E > 0$, we generate a random number between the range $(0, 1)$: $P' = 0.98818$. The probability of accepting the new point (P) is $e^{-(320.44/1.9531)} \cong 0$. Since $P' > P$, the new point is rejected. We set iteration counter $i = 10$ and proceed to step 4. Hence, after the 10th generation, new point $X^{(10)}$ remains at $X^{(9)}$.

Step 4: Lowering the Temperature as per the Cooling Schedule

The temperature (T) is lowered to $0.5 \times 1.9531 = 0.97656$. Since $T < T_{min}$, we terminate.

TABLE 11.1

A Summary of Search Points for First 10
Iterations

Iteration	Values of the Variables
$i = 0$	$X^{(0)} = (0, 0)$
$i = 1$	$X^{(1)} = (-7.337, 0.53718)$
$i = 2$	$X^{(2)} = (-7.337, 0.53718)$
$i = 3$	$X^{(3)} = (-8.2649, 0.6315)$
$i = 4$	$X^{(4)} = (-8.2649, 0.6315)$
$i = 5$	$X^{(5)} = (-8.9895, 1.6117)$
$i = 6$	$X^{(6)} = (-5.4466, 2.3564)$
$i = 7$	$X^{(7)} = (-5.4466, 2.3564)$
$i = 8$	$X^{(8)} = (-3.2893, 5.386)$
$i = 9$	$X^{(9)} = (-1.4429, 2.6077)$
$i = 10$	$X^{(10)} = (-1.4429, 2.6077)$

Table 11.1 gives a summary of the search points after each iteration. The progress of the search starting from the point $(0, 0)$ up to the first 10 iterations is displayed in Figure 11.9.

The MATLAB® coding for solving the aforesaid optimization problem given in Equation 11.3 using the simulated annealing method is presented in Section 11.5.

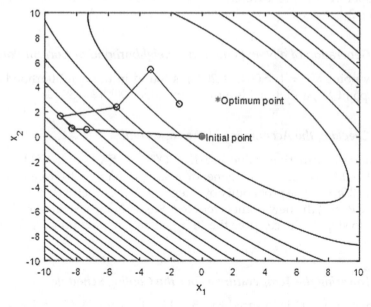

FIGURE 11.9

The progress of the search starting from the point $(0, 0)$ up to the first 10 iterations.

11.4 Application of Simulated Annealing in Textiles

Das et al. (2012, 2013) used the simulated algorithm for engineering design of woven fabrics with minimum manufacturing cost and requisite quality. The optimization problem was formulated to minimize the fabric areal density (GSM) subjected to the inequality constraints composed of fabric tensile, bending, and shear properties, and an equality constraint of crimp balance. The optimization problem was solved using simulated annealing to find out the optimum values of governing parameters such as yarn counts, crimps, and threads spacing for the production of lightweight, medium weight, and heavyweight cotton fabrics. Das et al. (2014) also made an attempt for optimal product design in the textile spinning industry using simulated annealing.

Heckmann and Lengauer (1995) used a simulated annealing approach to the nesting problem in the textile manufacturing industry, which is basically a problem of placing a set of irregularly shaped pieces (called stencils) on a rectangular surface, such that no stencils overlap and the trim loss produced when cutting out the stencils is minimized. They pointed out that the simulated annealing algorithm has high performance and is easy to implement.

11.4.1 Application of Simulated Annealing in Yarn Engineering

Das and Ghosh (2015) used a simulated annealing algorithm for engineering design of cotton yarn with predefined strength by optimal selection of raw material and process parameters. The mechanistic model of yarn strength developed by Zurek (1975), Zurek et al. (1987), and Frydrych (1992) has been used for formulation of the optimization problem, which is discussed as follows.

The metric twist factor α_m is expressed as

$$\alpha_m = t\sqrt{\frac{Tt_y}{1,000}} \tag{11.7}$$

where t is the yarn twist/meter, and Tt_y is the yarn linear density (tex).

The density of yarn ρ_y and α_m are related by the following expression (Barella, 1950):

$$\rho_y = 560 + 2.8\alpha_m \tag{11.8}$$

The nominal twist parameter is given as

$$g = 2\pi Rt = \sqrt{\frac{125.7}{\rho_y}\left(\frac{\alpha_m}{100}\right)} \tag{11.9}$$

where R is the yarn radius.

The equivalent fiber diameter is written as

$$d_f = \sqrt{\frac{4Tt_f}{\pi \rho_f}}$$
(11.10)

where Tt_f is linear density of a fiber (mg/m), and ρ_f is the fiber density (Kg/m³); for cotton $\rho_f = 1520$ Kg/m³.

The yarn diameter (in mm) can be expressed as

$$d_y = \sqrt{\frac{4Tt_y}{\pi \rho_y}}$$
(11.11)

The reduced twist parameter becomes

$$g_c = 2\pi \left(R - \frac{d_f}{4} \right) t = g \left(1 - \frac{1}{2} \frac{d_f}{d_y} \right)$$
(11.12)

The contraction factor is given by

$$s = \frac{\ln \sqrt{1 + g_c^2}}{\sqrt{1 + g_c^2} - 1}$$
(11.13)

The coefficient of the yarn diameter contraction can be calculated according to the following formula:

$$u = 1.13 - \frac{0.0265}{g_c} - 0.12 \sqrt[4]{100 a_h}$$
(11.14)

Since a_h is unknown, it is first assumed that a_h is fiber breaking strain a_f and the value of u is calculated. Knowing the approximated value of u, the strain of the breaking zone of the yarn is found from the following relationship:

$$a_h = \sqrt{\left(\frac{1 + a_h}{s} \right)^2 - u^2 g_c^2} - 1$$
(11.15)

Then a new value of u is calculated, and again a new value of a_h is found. This process is continued until the calculated values of u do not change.

The parameter of change of the fiber axis shape is written as

$$k = \frac{1 + a_h}{u}$$
(11.16)

The critical value of the twist parameter is obtained as

$$g_r = \frac{g}{k}\left(1 - \frac{2}{u}\frac{d_f}{d_y}\right) \tag{11.17}$$

The radius of the fiber axis curvature in the external layer of yarn is given by

$$P = (Ru - d_f)\frac{1 + g_r^2}{g_r^2} \tag{11.18}$$

The z is a parameter that represents the change in yarn strength because the fibers are constrained from moving freely in the yarn. The numerical value z can be calculated from the formula as follows:

$$f(z) = \frac{4P}{\mu(c\eta l_f - \lambda)} = \frac{1 - z}{(2z + 1)\ln\left(1 + \dfrac{1}{2z}\right) - 1} \tag{11.19}$$

where μ is the coefficient of friction between fibers; for cotton $\mu = 0.27$, c is a coefficient depending on the spinning system; $c = 1.1$ for combed yarn, η is the relative extent of fibers in yarn; $\eta = 0.9$ for combed yarn, l_f is the mean length of fibers, and λ is the length of fiber ends outside the yarn; it is assumed that $\lambda = 0.2$ mm.

The fracture zone length is expressed as

$$l_h = \frac{2P}{\mu}\ln\left(1 + \frac{1}{2z}\right) \tag{11.20}$$

The ratio of the specimen length y to the length of fracture zone l_h is

$$q = \frac{y}{l_h} \tag{11.21}$$

The parameter C depends on twist and the change in fiber axis shape while the yarn was being strained:

$$C = \frac{2k}{g^2}\ln\frac{\left[1.0253g^2 + 1\right]^{0.5} + \left[1.0253g^2 + k^2\right]^{0.5}}{\left[0.0253g^2 + 1\right]^{0.5} + \left[0.0253g^2 + k^2\right]^{0.5}} \tag{11.22}$$

The yarn stress at zero gauge length Q_h is calculated from the equation

$$Q_h = zQ(\varepsilon)C \tag{11.23}$$

where $Q(\varepsilon)$ is the breaking stress of the fiber.

The strength of a y mm long yarn sample Q_y is obtained by

$$Q_y = Q_h \left[1 - 3.64 v_{Fh} \left(1 - q^{-\frac{1}{7}}\right)\right] \tag{11.24}$$

where v_{Fh} is the variation coefficient of the breaking force in the length of the fracture zone, which can be approximately determined according to the relationship

$$v_{Fh} = \beta \sqrt{\frac{Tt_y}{Tt_f}} \tag{11.25}$$

where β is 1.35 for a combed spinning system. Here yarn specimen length y is taken as 500 mm.

The following optimization problem was formulated for 20 tex combed cotton yarn:

$$
\begin{aligned}
\text{Minimize} \quad & (Q_y - T_y)^2 \\
\text{Subject to} \quad & t^L \leq t \leq t^U \\
& Tt_f^L \leq Tt_f \leq Tt_f^U \\
& a_f^L \leq a_f \leq a_f^U \\
& Q(\varepsilon)^L \leq Q(\varepsilon) \leq Q(\varepsilon)^U \\
& l_f^L \leq l_f \leq l_f^U
\end{aligned}
\tag{11.26}
$$

where Q_y is the predicted yarn strength, T_y is the target yarn strength, and the superscripts L and U refer to the values of lower and upper boundaries of the variables, respectively. The target value of yarn strength T_y was set to 20 gf/tex.

Simulated annealing was used to solve the optimization problem of Equation 11.26 in order to obtain the predefined yarn strength by searching the optimum values of cotton fiber parameters such as linear density Tt_f, breaking strain a_f, breaking stress $Q(\varepsilon)$, and mean length l_f, as well as yarn twist/meter t. The values of initial temperature T, minimum temperature T_{min}, cooling rate C_T, and number of iterations at each temperature n for the simulated annealing algorithm were set to 1,000; 0.0001; 0.9; and 500, respectively.

Table 11.2 shows the lower and upper boundaries of the variables. Table 11.3 illustrates the optimum combination of fiber properties and twist/meter for the production of 20 tex yarns with target strength of 20 gf/tex. The value of yarn strength with the optimized parameters was obtained as 19.97 gf/tex.

TABLE 11.2

Boundary of Constraints

Controlling Parameters	Lower Boundary	Upper Boundary
Twist/meter (m^{-1})	650	700
Fiber linear density (tex)	0.120	0.150
Fiber breaking strain (%)	6.0	7.0
Average fiber length (mm)	25	33
Fiber breaking stress (gf/tex)	28	35

Source: Das, S. and Ghosh, A. 2015. *Fibres and Textiles in Eastern Europe*, 23(3), 51–53.

TABLE 11.3

Optimized Value of Constraints

Controlling Parameters	Optimized Values
Twist/meter	681.5
Fiber linear density (tex)	0.135
Fiber breaking strain (%)	6.75
Average fiber length	32.7
Fiber breaking strength (gf/tex)	34.3

Source: Das, S. and Ghosh, A. 2015. *Fibres and Textiles in Eastern Europe*, 23(3), 51–53.

11.5 MATLAB® Coding

MATLAB® Coding for Simulated Annealing

```
clc
clear all
close all
format short g
% Step-1: Initialization
x = [0  0];
vlb = [-10 -10];
vub = [10 10];
T = 1000;
Ct = 0.9;
T_min = 1e-6;
max_n = 1000;
i =0;
% Step-2: Generation of a new point in the neighborhood of
current point
sigma = [vub-vlb]/6;
while T > T_min
```

```
        for n = 1: max_n
            new_x = random('norm', x, sigma);
            if ((vlb <= new_x) & (new_x <= vub))
                fx = (x(1) + 2*x(2) -7).^2 + (2*x(1) + x(2)
-5).^2;
                fx_new = (new_x(1) + 2*new_x(2) -
7).^2+(2*new_x(1) + new_x(2) - 5).^2;
                % Step-3: Checking the acceptance of the new point
                del_E = (fx_new - fx);
                    if del_E < 0
                        x = new_x;
                        else r = rand;
                        if r < = exp(-del_E / T)
                            x = new_x;
                        end
                    end
                end
                i = i + 1;
            end
            % Step-4: Lowering the temperature as per the cooling
            schedule
            T = T*Ct;
end
i
x
f_val = (x(1) + 2*x(2) -7).^2 + (2*x(1) + x(2) -5).^2
```

The MATLAB® code for simulated annealing was executed to solve the optimization problem given in Equation 11.3. We have set $X^{(0)} = (0, 0)$, $T = 1,000$, $C_T = 0.9$, $T_{min} = 10^{-6}$, and $n = 1,000$. The temperature is reduced in steps, which has a geometric progression as follows:

$$1,000, 1,000 \times 0.9, 1,000 \times 0.9^2, \dots , 1,000 \times 0.9^{(s-1)}$$

where s is the number of steps associated with temperature reduction. It can be seen that just after 197 steps, the temperature becomes less than 10^{-6}. As 1,000 iterations have been performed at a particular temperature, the program terminates after $197 \times 1,000 = 197,000$ iterations. Figure 11.10 shows the progress in reducing the value of objective function over 197 steps, whereas Figure 11.11 depicts the same over 197,000 iterations. The algorithm eventually converges to the point (1.002, 2.994), which is very close to the actual optimum point (1, 3). The progress of the algorithm starting from the initial point to the end point is illustrated in Figure 11.12. It is clearly evident from Figure 11.12 that the search covers the whole space of variables; however, it ultimately concentrates in the neighborhood of the optimum point.

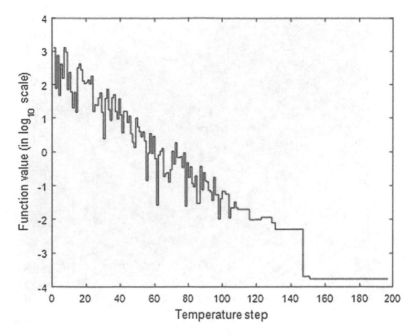

FIGURE 11.10
The progress of function value reduction with temperature step.

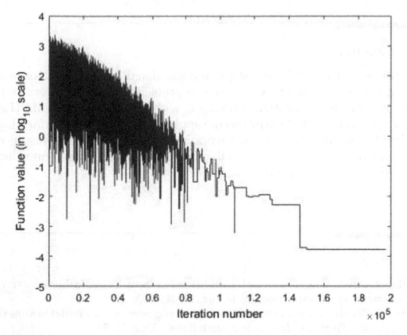

FIGURE 11.11
The progress of function value reduction with iteration number.

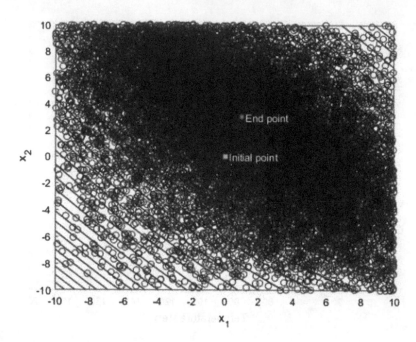

FIGURE 11.12
The progress of the search starting from the initial point to the end point.

11.6 Summary

An overview of the different steps of a simulated annealing algorithm is presented in this chapter. A worked-out problem brings out every step of the simulated annealing algorithm along with its coding in the MATLAB® language. A survey of literature on the application of simulated annealing in textiles shows only little reported work is available. Finally, the results of a study carried out by the author on the application of simulated annealing in a yarn engineering problem are discussed.

References

Barella, A. 1950. Law of critical yarn diameter and twist influence on yarn characteristics. *Textile Research Journal*, 20, 249–253.

Das, S. and Ghosh, A. 2015. Cotton fibre-to-yarn engineering: A simulated annealing approach. *Fibres and Textiles in Eastern Europe*, 23(3), 51–53.

Das, S., Ghosh, A. and Banerjee, D. 2012. Selection of parameters for engineering design of woven fabrics using simulated annealing. *International Conference on Innovations in Engineering and Technology for Sustainable Development*, Sathyamangalam, India, September 3–5.

Das, S., Ghosh, A. and Banerjee, D. 2013. Engineering design of woven fabrics using non-traditional optimization methods: A comparative study. *Fibers and Polymers*, 14(9), 1562–1567.

Das, S., Ghosh, A. and Saha, B. 2014. Optimal product design of textile spinning industry using simulated annealing. *Fourth International Conference on Soft Computing for Problem Solving*, Silchar, India, December 27–29, 315–323.

Deb, K. 2005. *Optimization for Engineering Design: Algorithms and Examples*, Prentice Hall, New Delhi, India.

Frydrych, I. 1992. A new approach for predicting strength properties of yarn. *Textile Research Journal*, 62(6), 340–348.

Heckmann, R. and Lengauer, T. 1995. A simulated annealing approach to the nesting problem in the textile manufacturing industry. *Annals of Operations Research*, 57(1), 103–133.

Kirkpatrick, S., Gelatt, C. D. and Vecchi, M. P. 1983. Optimization by simulated annealing. *Science*, 220, 671–680.

Metropolis, N., Rosenbluth, A., Rosenbluth, M., Teller, A. and Teller, E. 1953. Equation of state calculations by fast computing machines. *Journal of Chemical Physics*, 21, 1087–1092.

Zurek, W. 1975. *The Structure of Yarn*, Foreign Scientific Publications Department of the National Center for Scientific Technical and Economic Information, Warsaw, Poland.

Zurek, W., Frydrych, I. and Zakrzewski, S. 1987. A method of predicting the strength and breaking strain of cotton yarn. *Textile Research Journal*, 57, 439–444.

12

Multiobjective Optimization

12.1 Introduction

Most of the real-world problems involve more than one objective. Whenever an optimization problem involves two or more objectives, the computational complexity increases, and the task of finding optimum solutions of such problems is known as *multiobjective optimization*. The optimization solution of such problems may be either one or more depending on the type of objective. Whenever the objectives are nonconflicting in nature, a single optimum solution may be possible. But, when the objectives are conflicting in nature, more solutions may be possible, and a trade-off is then required. A solution that is extreme with respect to one objective needs to compromise with other objectives, and vice versa. There exist various classical and ideal optimization techniques that deal with such complex optimization problems. Classical techniques like *goal programming* and *desirability function*, and nonclassical, unorthodox search optimization processes like *evolutionary algorithm* are often used to solve multiobjective optimization problems. This chapter presents a detailed application of goal programming, desirability function, and evolutionary algorithm to deal with multioptimization problems in textile manufacturing.

12.2 Goal Programming

Goal programming (GP) is a classical optimization method to solve multiobjective optimization problems. It is an extension of the linear programming (LP) method that, unlike LP that has only one linear objective function, deals with multiple linear objectives (Panneerselvam 2002; Vohra 2010). Often, a manufacturing problem demands the optimization of multiple objectives. In such a multiple objective manufacturing problem, LP chooses only one of them as the objective function and considers others as constraints. Therefore, the solution of an LP problem ought to satisfy all of the constraints prior to optimization of the objective function. In a GP model, each objective or constraint is viewed as a "goal," and a target is set for each goal. In a real manufacturing scenario, these goals are often in conflict in nature and hence are unlikely to be satisfied simultaneously. Sometimes, an establishment of hierarchy of importance of the goals is also required. Lower-order goals in this hierarchy are considered only after higher-order goals are satisfied. After setting the target for the goals, each goal is converted into an equality constraint by assigning *deviation variables* or *slack variables* d_i^+ and d_i^-. The deviational variables specify the amounts a targeted goal i is overachieved or underachieved, respectively. An objective function is then developed which contains these deviational variables. Differential weights, if necessary, may be assigned to deviational variables at the same priority level. A solution is then attempted to minimize the deviational variables in order of priority. A general GP model is as follows:

Minimize $\quad Z = \displaystyle\sum_{i}^{n} w_i\left(d_i^+ + d_i^-\right)$

Subject to $\quad \displaystyle\sum_{i=1}^{k} a_{ij}x_j + d_i^- - d_i^+ = b_i \ (i = 1, 2, \ldots k)$

$$\begin{pmatrix} \text{Value of the} \\ \text{objective} \end{pmatrix} + \begin{pmatrix} \text{Amount below} \\ \text{the goal} \end{pmatrix} - \begin{pmatrix} \text{Amount above} \\ \text{the goal} \end{pmatrix} = \text{Goal}$$

$$x_j, d_i^-, d_i^+ \geq 0 \text{ for all } i, j$$

(12.1)

where x_j is decision variable j, w_i is the priority attached to goal i, d_i^- is underachievement of goal i, and d_i^+ is overachievement of goal i.

12.2.1 Goal Programming with One Goal

Though GP is intended to deal with problems with multiple goals, a problem with a single goal is explained initially to understand the basic concept of GP.

EXAMPLE 12.1

Let us consider two styles of apparels *Style A* and *Style B* that are being manufactured in an export house. Both styles use the same fabric and button. *Style A* uses 3 meters of fabric and 5 buttons, whereas *Style B* uses 2 meters of fabric and 10 buttons. The company has total weekly availability of 1,400 meters of fabric and 2,800 buttons. The total man-minutes available in the company per week is 16,800. It takes 40 and 30 man-minutes to manufacture one *Style A* and *Style B* apparel, respectively. The profits per unit of *Style A* and *Style B* apparel are Rs. 50 and Rs. 40 respectively. The problem is as follows:

$$\text{Maximize the profit} \quad Z = 50x_1 + 40x_2$$
$$\text{Subject to} \quad 3x_1 + 2x_2 \leq 1,400$$
$$5x_1 + 10x_2 \leq 2,800 \qquad (12.2)$$
$$40x_1 + 30x_2 \leq 16,800$$
$$x_1, x_2 \geq 0$$

where x_1 is the number of apparels in *Style A* and x_2 is the number of apparels in *Style B*.

Now, assume the production manager has set a target to achieve a profit of Rs. 21,500. Since it is not known whether the goal will be exactly met, overachieved, or underachieved, two deviational variables d^+ and d^- are introduced. The deviational variables d^+ and d^- denote overachievement and underachievement, respectively. Therefore, goal constraint for the above problem can be expressed as

$$50x_1 + 40x_2 + d^- - d^+ = 21,500 \qquad (12.3)$$

It is obvious from Equation 12.3 that when the goal is met exactly, the deviational variables d^+ and d^- would be equal to zero. In case the goal is overachieved, d^- would be zero and d^+ would be positive. Again, in case of underachievement of the goal, d^+ would be zero and d^- would be positive.

In this problem, there are three constraints and one converted goal constraint. To find the optimal solution, this problem is converted into a standard minimization or maximization problem using the concept of goal function. As the profit should be as close as possible to target Rs. 21,500, d^-, that is, underachievement, should be minimized. Accordingly, the problem becomes

$$\text{Minimize} \quad Z = d^-$$
$$\text{Subject to} \quad 3x_1 + 2x_2 \leq 1,400$$
$$5x_1 + 10x_2 \leq 2,800$$
$$40x_1 + 30x_2 \leq 16,800 \qquad (12.4)$$
$$50x_1 + 40x_2 + d^- - d^+ = 21,500$$
$$x_1, x_2, d^-, d^+ \geq 0$$

This problem may be solved using the LP problem, and the solutions are $x_1 = 336$, $x_2 = 112$, $d^- = 220$, $d^+ = 0$, and profit = Rs. 21,280. The solutions show that the profit goal is underachieved by Rs. 220, and the maximum profit is Rs. 21,280.

The MATLAB® code for solving the aforesaid single goal optimization problem is given in Section 12.5.

12.2.2 Goal Programming with Multiple Goals

Often manufacturing processes have more than one goal and such problems can be dealt with multiple GP. Similar to single GP, multiple GP sets specific targets which may be either met or underachieved or overachieved. Based on how these goals are, multiple GP is categorized into non-preemptive and preemptive. Non-preemptive GPs are those where the goals are of comparable importance, that is, there is no prioritization of goals. In case of preemptive GP, the goals have a hierarchy of importance. Therefore, goal with utmost importance receives attention before others which have relatively less importance. Hence, goals are considered one after another (and not simultaneously) based on their relative importance.

12.2.2.1 Non-Preemptive Goal Programming

As mentioned earlier, all the goals in a non-preemptive GP have equal importance. Initially, the goal for each objective is specified and then the sum of the deviations of these objectives from their respective goals is minimized. The general model of a non-preemptive GP is given below:

$$\text{Minimize} \quad \sum_{i=1}^{n}\left(d_i^+ + d_i^-\right)$$

$$\text{Subject to} \quad \sum_{i=1}^{k} a_{ij}x_j + d_i^- - d_i^+ = b_i \ (i = 1,2,\ldots k) \tag{12.5}$$

$$x_j, d_i^-, d_i^+ \geq 0 \text{ for all } i, j$$

where x_j is decision variable, d_i^- is underachievement of goal i and d_i^+ is overachievement of goal i.

An example of non-preemptive GP has been illustrated in Example 12.2.

EXAMPLE 12.2

Let us consider the same problem explained before in Example 12.1, where two types of apparels namely *Style A* and *Style B* have been manufactured in an export house. Let us further assume that the export house got a new assignment to produce two new type of apparels namely *Style C* and *Style D* which are replacements of *Style A* and *Style B* respectively. The weekly availability of fabric (in meters), buttons and man minutes are same as earlier, *that is*, 1,400, 2,800 and 16,800. The unit profit for *Style C* and *Style D* are same, that is, Rs. 30. The amount of fabric required for *Style C* and *Style D* are 2.5 meters and 2 meters respectively. *Style C* requires 5 buttons whereas *Style D* requires 10 buttons. The man minutes required for *Style C* and *Style D* are same as mentioned in the earlier problem, that is, 40 and 30 respectively. The production manager has set a profit target of Rs. 21,500 but not reluctant to give any overtime to the workers.

Now, in the above problem, there are two goals and four constraints. To find the optimal solution, the above problem is converted into standard minimization problem using the concept of goal function. As the production manager has set a target of profit as close as possible to Rs. 21,500 without any extra overtime to the workers, therefore it is required to minimize the underachievement of profit and the overachievement of total available man minutes. Thus, the sum of the deviations of these objectives from their respective goals (i.e., summation of underachievement and overachievement) is minimized. Thus, the problem can be formulized as,

$$
\begin{aligned}
\text{Minimize} \quad & Z = d_1^- + d_2^+ \\
\text{Subject to} \quad & 2.5x_1 + 2x_2 \leq 1,400 \\
& 5x_1 + 10x_2 \leq 2,800 \\
& 30x_1 + 30x_2 + d_1^- - d_1^+ = 21,500 \\
& 40x_1 + 30x_2 + d_2^- - d_2^+ = 16,800 \\
& x_1, x_2, d_1^-, d_1^+, d_2^-, d_2^+ \geq 0
\end{aligned}
\tag{12.6}
$$

This problem is solved using *LPP* and the solutions are $x_1 = 336$, $x_2 = 112$, $d_1^- = 8,060$, $d_1^+ = 0$, $d_2^+ = 0$, $d_2^- = 0$, profit = Rs. 13,440 and utilization of man minutes = 16,800. The solutions show that while the profit goal is underachieved by Rs. 8,060 and the maximum profit is Rs. 13,440, the utilization of man minute is perfectly met as per the target set.

The MATLAB® coding for solving the aforesaid optimization problem with multiple goals discussed in Example 12.2 is given in Section 12.5.

12.2.2.2 Goal Programming with Differential Weighting

In the last example (i.e., Example 12.2), no preference factor or weight was assigned for either under achievement or over achievement of goals (two goals were considered; profit and man minutes). It was assumed that unit deviation (under or over achievement) of one goal is same to that of other goal which is actually not. For an example, a deviation of Rs. 10 profit with a target of Rs. 21,500 would mean a deviation of $(10 \times 100)/21,500 = 0.05\%$ while a deviation of 5 man minutes with target of 16,800 man minute would imply deviation of 0.06%. Though in this instance, the difference shown in percentage is small due to small magnitude difference of two goals, a problem associated with large magnitude difference produces problem. Therefore, there is a need to assign preference vectors or weights against each deviation (under and over achievement) in the objective function. If we suppose that the percent deviations are of interest to the production manager, then the objective function can be restated as below:

$$
\text{Minimize} \quad Z = \frac{1}{21,500} d_1^- + \frac{1}{16,800} d_2^+
\tag{12.7}
$$

Again, if the decision maker feels that maintaining man minute is thrice as important as profit goal; the Equation 12.7 can be rewritten as:

$$
\text{Minimize} \quad Z = \frac{1}{21,500} d_1^- + \frac{3}{16,800} d_2^+
\tag{12.8}
$$

With the revised objective function and same set of constraints, the solution to the modified problem can be calculated as discussed earlier.

12.2.2.3 Preemptive Goal Programming

In preemptive GP, different goals have a hierarchy of priority level. Therefore, in a manufacturing process, the goals which seem more significant are given priorities to achieve their targets followed by other goals with lower priority.

Initially the goals are ranked in a hierarchical order from the goal with prime importance (say, goal 1) to the goal with least importance (say, goal n). Then the solution is obtained by optimizing the first priority goal/goals without considering the lower priority goals. Next, the optimal solution is obtained for the next priority goal keeping the optimal value of first priority goal as constraint function. Henceforth, the next priority goals are optimized successively by adding the constraint function of the earlier goals. The procedure continues until all priorities are considered. An example of preemptive GP is shown in Example 12.3.

EXAMPLE 12.3

An apparel manufacturing unit produces two various types of garments, namely X and Y. Each garment must be processed through two different departments, namely A and B. Department A has a capacity of total 48 man-hours per day and that for department B is 64 man-hours per day. Unit product of X requires 2 man-hours and 3 man-hours, whereas unit product of Y requires 3 man-hours and 4 man-hours in department A and B respectively. The marketing manager has prioritized the following goals depending on demand in the market.

Goal 1: Minimum number of Y to be produced daily $= 8$
Goal 2: Minimum number of X to be produced per day $= 6$
Goal 3: Total production per day $= 15$ garments

Let us assume x_1 and x_2 are number of garments A and B produced respectively, and d_i^- and d_i^+ underachievement and overachievement associated with goal i. The *GP* model for Goal 1 is stated as follow:

$$
\begin{aligned}
\text{Minimize} \quad & Z = d_1^- \\
\text{Subject to} \quad & 2x_1 + 3x_2 \le 48 \\
& 3x_1 + 4x_2 \le 64 \\
& x_2 + d_1^- - d_1^+ = 8 \\
& x_1 + d_2^- - d_2^+ = 6 \\
& x_1 + x_2 + d_3^- - d_3^+ = 15 \\
& x_1, x_2, d_1^-, d_1^+, d_2^-, d_2^+, d_3^-, d_3^+ \ge 0
\end{aligned}
\tag{12.9}
$$

This part of the problem is solved using *LPP* and the solutions are $x_1 = 0, x_2 = 8, d_2^- = 6$ and $d_3^- = 7$, with all other variables 0. Goal 1 is fulfilled but Goal 2 and Goal 3 are not fulfilled.

The MATLAB® coding for solving the optimization problem given in Equation 12.9 is presented in Section 12.5.

Next, the target of Goal 2 is achieved. To achieve the target of Goal 2, a constraint (i.e., $d_1^- = 0$) is added to keep the first goal at the value obtained earlier. Therefore, the *GP* model would be,

$$
\begin{aligned}
\text{Minimize} \quad & Z = d_2^- \\
\text{Subject to} \quad & 2x_1 + 3x_2 \le 48 \\
& 3x_1 + 4x_2 \le 64 \\
& x_2 + d_1^- - d_1^+ = 8 \\
& x_1 + d_2^- - d_2^+ = 6 \\
& x_1 + x_2 + d_3^- - d_3^+ = 15 \\
& d_1^- = 0 \\
& x_1, x_2, d_1^-, d_1^+, d_2^-, d_2^+, d_3^-, d_3^+ \ge 0
\end{aligned}
\tag{12.10}
$$

This part of the problem is solved using *LPP* and the solutions are $x_1 = 6, x_2 = 8$ and $d_3^- = 1$, with all other variables 0. Goal 2 is met, Goal 1 is retained but Goal 3 is not fulfilled.

The MATLAB® coding for solving the aforesaid optimization problem in Equation 12.10 is given in Section 12.5.

Lastly, the target of Goal 3 is accomplished. To achieve the target of Goal 3, two constraints (i.e., $d_1^- = 0$ and $d_2^- = 0$) are added to keep the first goal and second goal at the value obtained earlier. Therefore, the *GP* model becomes,

$$
\begin{aligned}
\text{Minimize} \quad & Z = d_3^- \\
\text{Subject to} \quad & 2x_1 + 3x_2 \le 48 \\
& 3x_1 + 4x_2 \le 64 \\
& x_2 + d_1^- - d_1^+ = 8 \\
& x_1 + d_2^- - d_2^+ = 6 \\
& x_1 + x_2 + d_3^- - d_3^+ = 15 \\
& d_1^- = 0 \\
& d_2^- = 0 \\
& x_1, x_2, d_1^-, d_1^+, d_2^-, d_2^+, d_3^-, d_3^+ \ge 0
\end{aligned}
\tag{12.11}
$$

The problem is solved using *LPP* and the solutions are $x_1 = 6$, $x_2 = 9$. All deviational variables are zero except $d_1^+ = 1$. All the goals are exactly met using the preemptive nature of GP.

The MATLAB® coding for solving the optimization problem given in Equation 12.11 is presented in Section 12.5.

12.3 Multiobjective Optimization Using Desirability Function

The *desirability function* is a traditional optimization technique in which two or more objectives can be optimized simultaneously. A desirability function (d_i) has been defined individually for each objective function with certain goals and boundaries. The desirability functions with different goals and boundaries are shown in Figure 12.1. Goals may be of three types, namely, maximize the objective, minimize the objective, and target the objective, and their expressions are given in Equations 12.12–12.14, respectively.

a. Maximize the objective

$$d_i = \left(\frac{Y_i - Y_{min}}{Y_{target} - Y_{min}} \right)^p \tag{12.12}$$

where $Y_{min} \leq Y_i \leq Y_{target}$
Minimize the objective

$$d_i = \left(\frac{Y_i - Y_{max}}{Y_{target} - Y_{max}} \right)^q \tag{12.13}$$

where $Y_{target} \leq Y_i \leq Y_{max}$
b. Target the objective

$$d_i = \left(\frac{Y_i - Y_{min}}{Y_{target} - Y_{min}} \right)^p \tag{12.14}$$

where $Y_{min} \leq Y_i \leq Y_{target}$
and

$$d_i = \left(\frac{Y_i - Y_{max}}{Y_{target} - Y_{max}} \right)^q$$

where $Y_{taregt} \leq Y_i \leq Y_{max}$

The exponents p and q in Equations 12.12–12.14 indicate the degree of importance to hit the target value. The desirability function approaches the target linearly when $p = q = 1$. The approach toward the target can be made in convex and concave paths when $p < 1, q < 1$ and $p > 1, q > 1$, respectively.

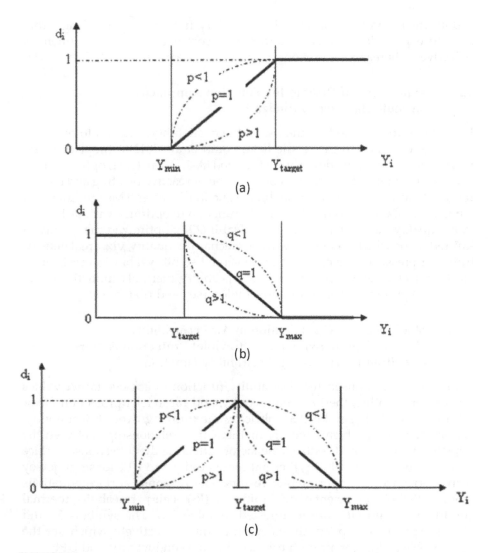

FIGURE 12.1
Desirability functions to (a) maximize, (b) minimize, or (c) reach a target value. (From Ghosh, A. et al. 2016. *Journal of Engineered Fibers and Fabrics*, 11(4), 20–28.)

The individual desirability functions (d_i) are combined to determine the overall desirability function for optimization. The overall desirability function is expressed in the following equation:

$$D = \sqrt[w]{\left(d_1^{w_1} \times d_2^{w_2} \times \cdots \times d_n^{w_n}\right)} \tag{12.15}$$

where $w = \Sigma w_i$, w_i is the weight of the ith objective and n is the number of objectives.

Both the individual and overall desirability functions have a range from zero to one. A desirability of zero means complete lack of fulfillment of objectives, whereas a value of one indicates complete fulfillment of objectives.

12.3.1 Application of Desirability Function Approach for Multiobjective Optimization

Few works are reported on the use of the desirability function to optimize multiple responses in the textile field. Souid et al. (2012) estimated denim fabric quality using the desirability function. Asim et al. (2011) optimized the process parameters for simultaneous fixation of reactive printing and crease-resistant finishing using the desirability function. Taieb and Msahli (2013) and Ghosh et al. (2016) used the desirability function to optimize various knitted fabric quality parameters. Taieb and Msahli (2013) optimized the quality of knitted fabric which composes various mechanical quality, viz. areal density, bursting pressure, extensibility, dimensional stability, abrasion resistance, etc. Ghosh et al. (2016) optimized air permeability, thermal conductivity, and ultraviolet protection factor for both single jersey and rib fabrics.

12.3.2 Multiobjective Optimization of Air Permeability, Thermal Conductivity, and Ultraviolet Protection Factor of Knitted Fabrics Using Desirability Function

Ghosh et al. (2016) used the desirability function to achieve target values for air permeability, thermal conductivity, and ultraviolet protection factor (UPF) for single jersey knitted fabric by maximizing overall desirability. As optimization problems require mathematical relationships between the objective functions and controllable factors affecting them, response surface equations of air permeability, thermal conductivity, and UPF for single jersey were derived from experimental results. The response surface equations along with their coefficient of determination (R^2) of air permeability, thermal conductivity, and UPF are manifested in Table 12.1. The symbols X_1 and X_4 correspond to loop length and yarn count, respectively, which are the controllable factors for air permeability, thermal conductivity, and UPF.

TABLE 12.1

Response Surface Equations for Single Jersey Fabric

Objective Functions	Response Surface Equations	Coefficient of Determination (R^2)
Air permeability ($cm^3/cm^2/s$)	$134.25 + 22.29X_1 + 111.49X_4 + 9.60X_1X_4 + 25.94X_4^2$	0.99
Thermal conductivity ($W/mK \times 10^{-3}$)	$38.85 - 0.78X_1 - 4.60X_4 + 1.16X_4^2$	0.95
UPF	$10.05 - 1.69X_1 - 7.69X_4 + 1.32X_1X_4 + 2.05X_4^2$	0.97

Source: Ghosh, A. et al. 2016. *Journal of Engineered Fibers and Fabrics*, 11(4), 20–28.

TABLE 12.2

Boundaries and Target Values of Different Objectives

Fabric Type	Objectives	Lower Limit	Target	Upper Limit
Single jersey	Air permeability (cm³/cm²/s)	36.01	80	303.57
	Thermal conductivity (W/mK×10⁻³)	34.63	40	45.39
	UPF	4.04	15	22.8

Source: Ghosh, A. et al. 2016. *Journal of Engineered Fibers and Fabrics*, 11(4), 20–28.

The minimum and maximum values for air permeability, thermal conductivity, and UPF were estimated from their corresponding response surface equations. Table 12.2 shows the target, lower and upper limits of air permeability, thermal conductivity, and UPF for single jersey fabric. To find the individual desirability for each objective, response equations as given in Equation 12.14 were used. In this study, the values of p and q were chosen

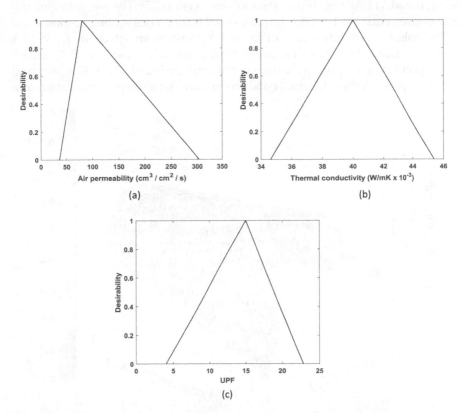

(a)

(b)

(c)

FIGURE 12.2

Individual desirability function diagram of (a) air permeability, (b) thermal conductivity, and (c) UPF for single jersey fabric. (From Ghosh, A. et al. 2016. *Journal of Engineered Fibers and Fabrics*, 11(4), 20–28.)

TABLE 12.3

Solutions of Optimization Problems for Single Jersey Fabric

Fabric Type	Response	Predicted Value of Response	Individual Desirability	Overall Desirability	Solution
Single jersey	Air permeability (cm³/cm²/s)	80.6	0.997	0.91	$X_1 = 6.6$ mm $X_4 = 6.7$ Ne
	Thermal conductivity (W/mK×10⁻³)	41.3	0.759		
	UPF	15	1		

Source: Ghosh, A. et al. 2016. *Journal of Engineered Fibers and Fabrics*, 11(4), 20–28.

as 1. The diagrams depicting individual desirability of each objective of single jersey fabric are depicted in Figure 12.2. The optimization problem of maximizing the overall desirability as expressed in Equation 12.15 was solved using the MATLAB® optimization toolbox (version 7.7). The weight value (w_i) was chosen equal to 1, which implies equal importance among responses.

The solutions of aforesaid optimization problem are given in Table 12.3. It is seen from the table that the overall desirability value is 0.91, which corresponds to the optimum values of 80.6 cm³/cm²/s, 41.3 W/mK $\times$ 10⁻³, and 15 for air permeability, thermal conductivity, and UPF, respectively, at 6.6 mm

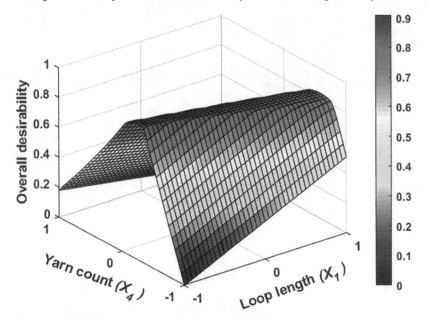

FIGURE 12.3

Overall desirability for single jersey fabrics (loop length and yarn count are in coded levels). (From Ghosh, A. et al. 2016. *Journal of Engineered Fibers and Fabrics*, 11(4), 20–28.)

loop length and 6.7 Ne yarn count. The values of individual desirability for air permeability, thermal conductivity, and UPF are 0.997, 0.759, and 1, respectively. It implies that the individual desirability of air permeability and UPF have been fulfilled, whereas for thermal conductivity, it is lacking to some extent. The surface plots of overall desirability function for single jersey are shown in Figure 12.3.

The MATLAB® coding for the aforesaid multiobjective optimization of air permeability, thermal conductivity, and UPF of single jersey fabric using the desirability approach is given in Section 12.5.

12.4 Multiobjective Optimization Using Evolutionary Algorithm

In most of the traditional methods of optimization of multiple objectives, the problems are made easier through transformation of the multiobjective problem into a single objective problem by using preference vectors. A composite objective function is usually formed by combining objective functions, each of which assigned some preference vectors or weightage. This procedure is known as preference-based multiobjective optimization. Based on experience, preference vectors were chosen, and then the composite objective function is optimized. When such composite objective function is optimized, only one solution may be obtained. Though the process is simpler, it is subjective and, hence, not ideal. A change of preference vector will give a different solution. It is unarguable that assigning a preference vector is highly subjective and not straightforward and creates an ambiguous situation. To understand better, the example cited in Section 12.4 may be considered where a preference-based optimization technique, i.e., desirability function approach, has been used. The overall desirability was maximized to obtain the optimal solutions. It is to be noted that the individual desirability of air permeability, thermal conductivity, and UPF were not considered. The overall desirability for the single jersey fabric was obtained as 0.91. It is evident from Table 12.3 that the optimum values for air permeability, thermal conductivity, and UPF for single jersey fabric were 80.6 cm³/cm²/s, 41.3 W/mK × 10^{-3}, and 15, respectively, and their individual desirability values were 0.997, 0.759, and 1, respectively. Now, the same overall desirability value can also be obtained if individual desirability values of air permeability, thermal conductivity, and UPF are 1, 0.997, and 0.759, respectively. The optimum values of objectives then become 80 cm³/cm²/s, 40.02 W/mK × 10^{-3}, and 16.88 for air permeability, thermal conductivity, and UPF, respectively. This creates a paradoxical situation. Furthermore, the desirability approach is a preference-based optimization process in which individual desirability depends on the preference vectors p and q as mentioned in Equations 12.12–12.14. Moreover, the outcome of the overall desirability depends on the preference vector w mentioned in Equation 12.15. To make the problem simpler and due to lack of exact information on correlations among the response types, preference vectors p, q, and w are considered as 1 in this multiobjective optimization problem. A change in any of these preference vectors will result in a different solution. Moreover, finding such a relative preference vector itself is highly subjective and not straightforward. It means that this approach of multiobjective optimization does not take into consideration the correlations among the response types and the variability of the predictions. In addition, in the desirability function approach, since all the objectives are combined together to form an objective (i.e., overall desirability), which is to be maximized, only one optimum solution is obtained. But, in an ideal multiobjective optimization problem, when the objective functions are conflicting in nature, different solutions may produce conflicting scenarios among different objectives. A solution that is better with respect to one objective (say air permeability) may require some compromise in another objective (say thermal conductivity). For example, in the earlier example of optimization using the desirability function, decision-making is involved to design knitted fabrics with optimal thermophysiological comfort with a desired level of ultraviolet (UV) protection. Let us consider two extreme hypothetical situations for single jersey fabrics having air permeability

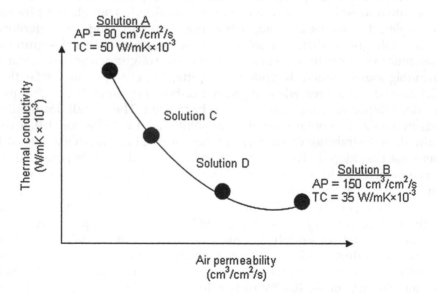

FIGURE 12.4
Hypothetical trade-off solutions are illustrated for decision-making problem of comfort of single jersey fabric.

values of 150 cm³/cm²/s (say, solution B) and 80 cm³/cm²/s (say, solution A), respectively, as shown in Figure 12.4.

If the air permeability is the only objective of this decision-making process, the optimum choice is 150 cm³/cm²/s. However, it is expected that a single jersey fabric with high air permeability is likely to have less thermal conductivity, and vice versa. Figure 12.4 illustrates that the single jersey fabric with higher air permeability (150 cm³/cm²/s) has a thermal conductivity of 35 W/mK × 10^{-3} (solution B). Solution A has lower air permeability (80 cm³/cm²/s) coupled with higher thermal conductivity 50 W/mK × 10^{-3}. Therefore, both the solutions as explained can be chosen depending on the choice of designer. Between these two extreme solutions, many other solutions (solutions C, D, etc.), are present where a trade-off between air permeability and thermal conductivity exists (Figure 12.4). It is observed that for any pair of solutions, one may be better than the other in terms of air permeability. However, this betterment comes from a compromise in terms of thermal conductivity. In multiobjective optimization, these multiple optimal solutions are termed as *Pareto optimal solutions* or *nondominated solutions* in which all solutions are equally important. For clarity, these optimal solutions are joined with a curve that is termed as *Pareto optimal front* (Deb, 2001). Depending on the requirement of the user, any one of the Pareto optimal solutions can be selected using high-level qualitative information. Unlike other classical methods, a set of trade-off solutions are made available using a scientific process, and a choice can be made from these sets of solutions. Hence, this ideal method of optimization is more methodical, unambiguous, and less subjective. Since the evolutionary algorithm finds multiple optimal solutions in a single simulation run, it becomes a unique technique in solving multiobjective optimization problems (Deb, 2010).

Genetic algorithms (GAs) are the nontraditional search and optimization methods based on genetics and natural selection. A GA mimics nature's evolutionary principles (Darwinian theory of survival of the fittest) to drive its search toward optimal solutions. A GA starts by randomly generating a population of initial solutions (represented by chromosomes). The initial population size depends on the complexity of optimization problem. Each individual solution (chromosome) is represented by a binary-coded string composed of 1 and 0s. Each gene of a chromosome is represented by each bit of the binary-coded string, and the length of the string is chosen based on the desired solution accuracy. At each iteration, the solutions of current population are tested for fitness. Based on the fitness information of initial solutions, good solutions are selected to form a new population (next generation) by an operator known as a *reproduction* or *selection* operator. It means that the reproduction operator produces duplicates of good solutions and eliminates bad solutions, while keeping the population size constant. This task of reproduction can be done by roulette wheel selection, tournament selection, proportionate selection, ranking selection

(Goldberg and Deb, 1991), etc. Reproduction or selection cannot introduce new solutions into a population. New solutions can be generated in a population by two operators, namely, *crossover* and *mutation*. A crossover operator exchanges the properties between two parent strings or solutions and results in two children strings (offsprings). The crossover operator is applied to only selected chromosomes, and its application is governed by a *crossover probability*, denoted by p_c. Too large p_c may prematurely destroy a high-quality solution, whereas too small p_c reduces the searching efficiency. Generally, p_c is chosen between 0.5 and 0.8. The mutation operator is used to bring a local change over the current solution by flipping bits of an individual's string at random. Thus, if a solution gets stuck at local minima, mutation may help it to come out from the local basin to jump into the global minima. The mutation operator is applied randomly with small probability, known as *mutation probability*, which is denoted by p_m. Mutation probability (p_m) is generally kept at a low value to avoid a random search. However, if p_m is too small, then a new gene segment cannot be induced. Generally, p_m is chosen between 0.001 and 0.1. Once the population of strings is modified using reproduction, crossover, and mutation operators, one iteration of a GA is complete. The GA runs until the termination criterion is reached. The termination criterion may be the maximum number of generations or a desired precision in the solution. Furthermore, in a GA, to avoid any loss of best solution, an *elite-preserving operator* is often recommended. The elite-preserving operator identifies elite strings (in terms of fitness) in a population and then directly copies them into the next iteration to ensure its presence. Deb (2001) has developed Non-dominated Sorting Genetic Algorithm II (NSGA-II) to optimize multiple objectives that are conflicting in nature.

The goal of the NSGA-II algorithm is to find a set of solutions that are as close as possible to the Pareto optimal front and as diverse as possible simultaneously. Except for the fitness assignment method, the basic structure of NSGA-II is similar to that of the GA (Sette and Lengenhove, 2002). The steps involved in this algorithm are briefly explained in the following sections (Goldberg, 1989; Srinivas and Deb, 1995; Deb et al., 2002; Skordos et al., 2002; Ghosh et al., 2013).

Step 1: Initialization of Random Binary Population

First a binary-coded population of size 2N is randomly initialized. Each individual of the population is represented by a binary string. The binary-coded strings are then converted into real values of variables using the following expression:

$$x_i = x_i^L + \left(\frac{x_i^U - x_i^L}{2^{l_{s_i}} - 1} \right) x_i^D$$

(12.16)

where x_i, x_i^L, x_i^U, and ls are the real value, lower limit, upper limit, and string length of the ith input parameter, respectively; and x_i^D is the decoded value of the ith parameter.

Step 2: Fast Nondominated Sorting

In this step, the population is sorted based on their nondomination levels. In this technique, two entities are calculated, the first is the domination count (n_i) that represents the number of solutions that dominate solution i, and the second is S_i that represents the number of solutions that are dominated by solution i. This is accomplished by comparing each solution with every other solution and checking whether the solution under consideration satisfies the rules as follows:

$$
\begin{array}{ll}
\text{objective } 1_i > \text{objective } 1_j & \text{and} \quad \text{objective } 2_i \geq \text{objective } 2_j \text{ or} \\
\text{objective } 1_i \geq \text{objective } 1_j & \text{and} \quad \text{objective } 2_i > \text{objective } 2_j, i \neq j
\end{array}
\tag{12.17}
$$

where objective 1_i and objective 1_j are fitness values of the first objectives for the ith and jth solutions, respectively. Similarly, objective 2_i and objective 2_j are fitness values of the second objectives for the ith and jth solutions, respectively.

If these rules are satisfied, then the solution j is dominated by solution i, or else it is nondominated. Thus, the whole population is divided into different ranks as shown in Figure 12.5. Ranks are defined as several fronts generated from the fast nondominated sorting technique such that Rank-1 solutions are better than Rank-2 solutions, and so on.

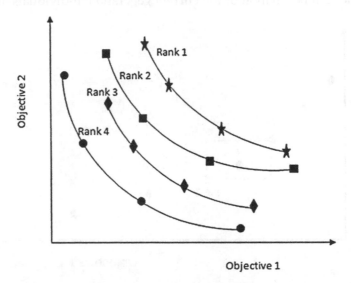

FIGURE 12.5
Sorting of population based on the members' nondomination levels.

Step 3: Crowding Distance

Once the populations are sorted, crowding distance is assigned to each individual belonging to each rank. This is because the individuals of next generations are selected based on rank and crowding distance. In order to calculate the crowding distance, the fitness levels of solutions belonging to a particular rank are sorted in descending order with respect to each objective. An infinite distance is assigned to the boundary solutions, that is, for the first and nth solutions, if n number of solutions belongs to a particular rank. This ensures that the individuals in the boundary will always be selected, and hence, it results in better spread among the solutions (Goldberg, 1989). For other solutions belonging to that rank, the crowding distances are initially assigned to zero. For $r = 2$ to $(n - 1)$ solutions, it is calculated by the following formula:

$$I(r)m = I(r)m + \frac{f_m(r-1) - f_m(r+1)}{f_m^{max} - f_m^{min}} \tag{12.18}$$

where $I(r)m$ is the crowding distance of the rth individual for the mth objective, $m = 1$ and 2, $f_m(r - 1)$ and $f_m(r + 1)$ are the values of the mth objective for $(r - 1)$th and $(r + 1)$th individuals, and f_m^{max} and f_m^{min} are the maximum and minimum values of the mth objective, respectively.

The crowding distance calculation is illustrated in Figure 12.6.

The 2N solutions are sorted based on their nondomination, and then crowding distances are calculated for all individuals belonging to a front. In order to form a population of the current generation, individuals are taken

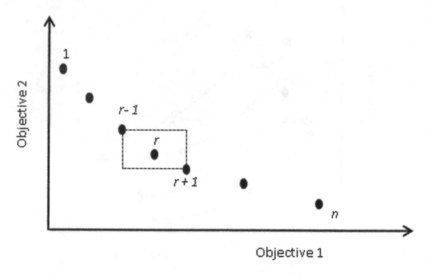

FIGURE 12.6
Crowding distance calculation for rth solution.

from the fronts subsequently unless it reaches the desired population number of N. The filling starts with the best nondominated front (Rank-1 solutions), then with the solutions of the second dominated front, followed by the third nondominated front, and so on. If by adding all individuals in a front the population exceeds N, then individuals are selected based on their crowding distance. In this process, once the number of population reaches N, we move to the next step.

Step 4: Crowded Tournament Selection

A crowded comparison operator compares two solutions and returns the winner of the tournament. A solution i wins a tournament with another solution j if any of the following conditions are true:

If solution i has a better rank than j

If they have the same rank but solution i has larger crowding distance than solution j

Tournaments are made to be played between N members of the parent population (P_t). The better member from each tournament is selected as part of a mating pool for crossover. Each of the members of the parent population is allowed to participate in two tournaments, and the better member is selected. It is unarguable that the best member in the population will win both the tournaments and will double up, whereas the worst member will lose both and will be eliminated.

Step 5: Crossover and Mutation

It is obvious that the operation of tournament selection copies good solutions, but it cannot create new solutions. New solutions are possible only by crossover and mutation operations. During crossover, two strings are randomly picked from the mating pool, and some portion of the strings is exchanged to create new strings. During crossover operation, a new population is created. After crossover, mutations of few strings are accomplished for further search of new solutions. After crossover and mutation, an offspring (Q_t) is thus created.

Step 6: Recombination and Selection

The offspring (Q_t) and parent population (P_t) are combined to form a $2N$ population, and selection of the best N is done by fast nondominated sorting and measuring crowding distance, as discussed before. Hence, the population of next-generation ($P_{(t+1)}$) is obtained. The elitism is ensured as the best population from offspring, and parent solutions are selected for the next generation. These steps are repeated until the maximum generation number is reached, which results in a Pareto optimal front. The schematic representation of NSGA-II is illustrated in Figure 12.7.

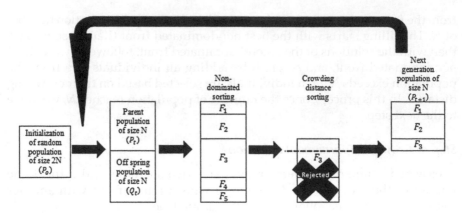

FIGURE 12.7
Schematic of NSGA-II.

In order to better understand the working of NSGA-II, a two-objective maximization problem is explained as follows:

$$\text{Maximize } f_1(x) = x_1 \tag{12.19}$$

$$\text{Maximize } f_2(x) = \frac{x_2}{x_1}$$
$$\text{subject to } 1 < x_1 < 10 \tag{12.20}$$
$$1 < x_2 < 100$$

It is obvious from this maximization problem that the two objectives, namely, $f_1(x)$ and $f_2(x)$, are conflicting in nature. An increase in the value of x_1 increases the value of $f_1(x)$ but decreases the value of $f_2(x)$, and vice versa. The steps of the hand calculation of the Pareto optimal solution for this problem are as follows.

Step 1: Initialization of Random Binary Population

An initial population (P_o) of size 2N has been selected randomly. For the sake of simplicity and better understanding, the population size taken in this example is $2N = 12$. Now, each member in the population is expressed by a binary-coded string that represents both the decision variables x_1 and x_2. Each decision variable is represented by an 8-bit binary substring, thereby making the overall string length equal to 16. The decoded values (D_1 and D_2) of these substrings are then calculated. Subsequently, the real values of x_1 and x_2 of all 12 strings are determined from their corresponding decoded values. The real values of x_1 and x_2 can be determined using the linear mapping rule given in Equation 12.16. For an example, the

TABLE 12.4

Coded, Uncoded, Real Values of Variables and Objective Values

String Number	Coded Values		Decoded Values		Real Values		Objective Function Values	
	Substring 1	Substring 2	D_1	D_2	x_1	x_2	$f_1(x)$	$f_2(x)$
1	11001001	01110101	201	117	8.0941	46.424	8.0941	5.7355
2	10110110	01100010	182	98	7.4235	39.047	7.4235	5.2599
3	11000000	10100101	192	165	7.7765	65.059	7.7765	8.3661
4	11001100	10001000	204	136	8.2	53.8	8.2	6.561
5	01101000	10001000	104	136	4.6706	53.8	4.6706	11.519
6	00110111	00011111	55	31	2.9412	13.035	2.9412	4.432
7	00011101	00011001	29	25	2.0235	10.706	2.0235	5.2907
8	11010110	11001100	214	204	8.5529	80.2	8.5529	9.3769
9	10010110	01110011	150	115	6.2941	45.647	6.2941	7.2523
10	00110110	01110111	54	119	2.9059	47.2	2.9059	16.243
11	10010101	01000000	149	64	6.2588	25.847	6.2588	4.1297
12	01111001	11010110	121	214	5.2706	84.082	5.2706	15.953

decoded value of the first substring (i.e 11001001) given in Table 12.4 is $1 \times 2^7 + 1 \times 2^6 + 0 \times 2^5 + 0 \times 2^4 + 1 \times 2^3 + 0 \times 2^2 + 0 \times 2^1 + 1 \times 2^0 = 201$. Thus, the corresponding real value of $x_1 = 1 + (10 - 1)/(2^8-1) \times 201 = 8.0941$. Similarly, the corresponding real value of x_2 is 46.424. Therefore, the corresponding values of the objective functions $f_1(x)$ and $f_2(x)$ are 8.0941 and 5.7355, respectively. The coded values; decoded values; real values of the variables x_1, x_2; and objective function values of the initial population (P_o) are shown in Table 12.4.

Step 2: Fast Nondominated Sorting

The population (P_o) is sorted based on its nondomination levels. In this step, each solution is compared with every other solution and checked whether the solution under consideration satisfies the rule mentioned in Equation 12.17. For an example, the objective function values of solution number 1 (string number 1) in Table 12.4 are compared with all other objective function values of 11 remaining solutions. It is obvious from the data in Table 12.4 that the solutions numbered 8, 10, and 12 dominate all other solutions but nondominate with respect to each other. Therefore, solutions numbered 8, 10, and 12 are placed in Rank 1. Thereby, all 12 solutions are divided into four different ranks, i.e., ranks 1–4, as shown in Table 12.5 and Figure 12.8. Ranks are defined as the fronts generated from the fast nondominated sorting technique. All of the solutions in the same front are nondominated in nature. The minimum front (i.e., front 1) has better solutions than the next higher front (i.e., front 2), solutions in front 2 have better solutions than front 3,

TABLE 12.5

Rank-Wise Solutions

Front	Solutions
Front 1	8, 10, 12
Front 2	3, 4, 5
Front 3	1, 9
Front 4	2, 6, 7, 11

and solutions in front 3 have better solutions than front 4. Out of 12 initial solutions, the best 6 solutions are selected for the next operation. The rank-wise solutions are given in Table 12.5. The fast nondominated sorting of 12 strings is illustrated in Figure 12.8.

In this example, the best 6 solutions out of a total of 12 solutions are directly chosen, as only 6 solutions in combination are laying in front 1 (3 solutions) and front 2 (3 solutions). If in case 5 solutions ought to be chosen, 3 solutions in front 1 will be selected along with 2 more solutions from front 2. Out of 3 solutions in front 2, the solutions that have higher crowding distance would be preferred for selection. The crowding distance for each solution in a front can be calculated using Equation 12.18.

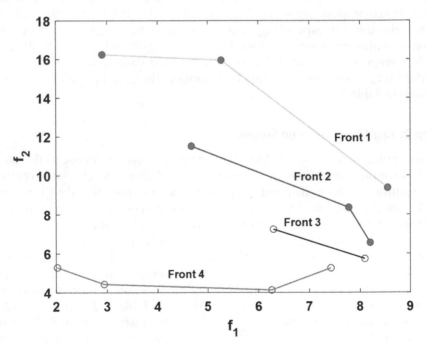

FIGURE 12.8
Fast nondominated sorting of 12 strings.

Step 3: Crowding Distance

The crowding distance is assigned to each individual solution belonging to each front using Equation 12.18. The individuals of the next generation are selected based on the front and crowding distance. In this example, a total of 6 solutions are sorted from a total of 12. Initially, all three solutions from front 1 are selected, and the remaining three are selected from front 2. As front 2 has three solutions, all of them are directly selected, making a total of six best solutions from front 1 and front 2. In case five solutions are required, two solutions from front 2 with higher crowding distance should be chosen. The hands-on calculation of crowding distance of the first best six solutions is as follows.

In this example, the number of solutions in front 1 is $l = 3$; $f_1^{max} = 9.9999$, $f_1^{min} = 1.0001$, $f_2^{max} = 99.9899$, and $f_2^{min} = 0.1$. We put $I(8)1 = I(10)1 = I(12)1 = 0$. Since solutions numbered 8 and 10 are boundary solutions for f_1, we put $I(8)1 = I(10) = \infty$. For the solution number 12, we get

$$I(12) = 0 + \frac{I(8)1 - I(10)1}{f_m^{max} - f_m^{min}} = 0 + \frac{8.5529 - 2.9059}{9.9999 - 1.0001} = 0.6275$$

Now, for the second objective function, we get

$$I(12) = 0.6275 + \frac{I(10)2 - I(8)2}{f_m^{max} - f_m^{min}} = 0.6275 + \frac{16.243 - 9.3769}{9.9899 - 0.1} = 0.6943$$

The overall crowded distances of the three solutions in rank 1 are as follows:

$$I(8) = \infty, I(10) = \infty \quad \text{and} \quad I(12) = 0.6943$$

In this example, the number of solutions in front 2 is $l = 3$; $f_1^{max} = 9.9999$, $f_1^{min} = 1.0001$, $f_2^{max} = 99.9899$, and $f_2^{min} = 0.1$. We put $I(3)1 = I(4)1 = I(5)1 = 0$. Since, solutions numbered 4 and 5 are boundary solutions for f_1, we put $I(4)1 = I(5) = \infty$. For solution number 3, we get

$$I(3) = 0 + \frac{I(4)1 - I(5)1}{f_m^{max} - f_m^{min}} = 0 + \frac{8.2 - 4.6706}{9.9999 - 1.0001} = 0.3922$$

Now, for the second objective function, we get

$$I(3) = 0.3922 + \frac{I(5)2 - I(4)2}{f_m^{max} - f_m^{min}} = 0.3922 + \frac{11.519 - 6.561}{9.9899 - 0.1} = 0.5013$$

The overall crowded distances of the three solutions in front 2 are

$$I(4) = \propto, I(5) = \propto, I(3) = 0.5013$$

The crowding distances of the solutions in front 1 and front 2 are shown in Table 12.6. The new population thus created is named the *parent population* (P_t).

Step 4: Crowded Tournament Selection

Tournaments are made played between six members of the parent population (P_t). The better member from each tournament is selected and is put in a mating pool for crossover. Each of the members of the parent population is allowed to participate in two tournaments. The tournaments played between the members of the parent population (each member getting two chances) and selection of the better member are illustrated in Figure 12.9. It is unarguable that the best member in the population will win both the tournaments and will double up, whereas the worst member will lose both and will be eliminated. In this example, the best members, that is, "10" and "8" have been doubled, whereas the worst member "3" is eliminated. (Both solutions 5 and 3 lie in front 2 and hence are nondominated in nature. As the crowding distance for 5 is more than 3, solution 5 is selected over solution 3.)

Step 5: Crossover and Mutation

It is obvious that the reproduction operation makes copies of good solutions but cannot create new solutions. In this example, worst solution 3 has been

TABLE 12.6

Fitness Assignment Procedure

Solution	x_1	x_2	f_1	f_2	f_1	f_2	Crowding Distance
		Front 1			Sorting in		
8	8.5529	80.2	8.5529	9.3769	First	Third	$\propto$
10	2.9059	47.2	2.9059	16.243	Third	First	$\propto$
12	5.2706	84.082	5.2706	15.953	Second	Second	0.6943
Solution	x_1	x_2	f_1	f_2	f_1	f_2	Distance
		Front 2			Sorting in		
3	7.7765	65.059	7.7765	8.3661	Second	Second	0.5013
4	8.2	53.8	8.2	6.561	First	Third	$\propto$
5	4.6706	53.8	4.6706	11.519	Third	first	$\propto$

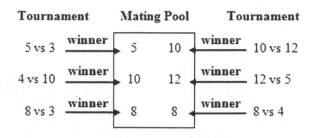

FIGURE 12.9
Tournaments played by six population members (each member getting two chances) and the winners.

eliminated; whereas, good solutions 10 and 12 have increased twofold. New solutions are possible only by crossover and mutation operations. During crossover, two strings are randomly picked from the mating pool, and some portion of the strings is exchanged to create two new strings. In this example, the crossover operation of the winners of the tournaments is shown in Table 12.7.

During crossover operation, a new population is created. After crossover, mutations of few strings are accomplished for the further search of new solutions. In this example, only one set of strings is mutated by changing the bits 1 to 0, and vice versa. An offspring (Q_t) is thus created. The mutation operation and offspring formation are shown in Table 12.8. The coded, uncoded, real values of variables and objective values of the offspring population are shown in Table 12.9.

Step 6: Recombination and Selection

The offspring (Q_t) and current population (P_t) are combined, and selection of the best six solutions is done by fast nondominated sorting and measuring crowding distance as previously discussed. Hence, the population of the next generation (P_{t+1}) is obtained. Elitism is ensured as the best population from offspring and parent solutions is selected for the next generation. The $2 \times 6 = 12$ solutions are sorted based on their nondomination, and then crowding distances are calculated for all individuals belonging to a front. In order to form a population of the current generation, individuals are taken from the fronts subsequently unless it reaches the desired population number of 6. These steps are repeated until the maximum generation number is reached, which results in a Pareto optimal front.

The decision variables space, objective space, Pareto optimal front, and objective space with Pareto optimal front of the previous problem are shown in Figures 12.10 through 12.13, respectively, considering an initial population size of 1,000. In this problem, the offspring have been generated

TABLE 12.7

Crossover Operation of Two Strings

Pair of Solutions	Parent Population (P_t)		Crossover (Y/N)	Population after Crossover	
	String	String		String	String
(8,10)	11010110	00110110	Yes	00010110	11010110
(5,10)	01101000	00110100	Yes	00110100	01101000
(12,8)	11010001	11001110	Yes	11010110	11001001

TABLE 12.8

Mutation Operations and Offspring Formation

Pair of Solutions	Population after Crossover		Mutation	Population after Mutation (Q_t)	
	String	String		String	String
(8,10)	00010110	11000101	First 11th bit	01011110	11000101
(5,10)	00110110	01101000	No	00111010	11000101
(12,8)	11011110	11011101	No	11011110	11000100

TABLE 12.9

Coded, Uncoded, Real Values of Variables and Objective Values of Offspring

String Number	Offspring (Q_t)		Uncoded Values x_1	x_2	Real Values of Variables x_1	x_2	Objective Values $f_1(x)$	$f_2(x)$
1	10010110	01011110	150	94	6.2941	37.4941	6.2941	5.9570
2	00110110	00111010	54	58	2.9059	23.5176	2.9059	8.0931
3	01110010	11011110	114	222	5.0235	87.1882	5.0235	17.3560
4	11110110	11000101	246	197	9.6824	77.4824	9.6824	8.0024
5	01101000	11000101	104	197	4.6706	77.4824	4.6706	16.5894
6	11011101	11000100	221	196	8.8	77.0941	8.8	8.7607

by a crossover probability of 0.8 and bitwise mutation with a probability of 0.001. The GA is run for 50 generations resulting in a Pareto optimal front. The Pareto optimal front shown in Figure 12.12 has 407 nondominated solutions. Since all of the solutions in the Pareto optimal front are better than the others, at least in terms of one objective, any one of them is acceptable, and the choice of solution depends on the consumer's requirement or high-level information. The MATLAB® coding for the aforesaid NSGA-II example is given in Section 12.5.

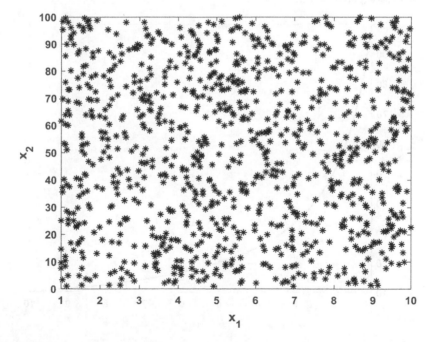

FIGURE 12.10
Decision variable space (population size = 1,000).

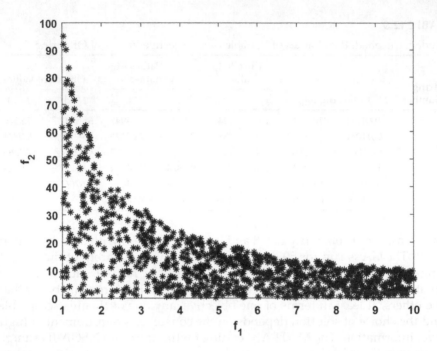

FIGURE 12.11
Objective space (population size = 1,000).

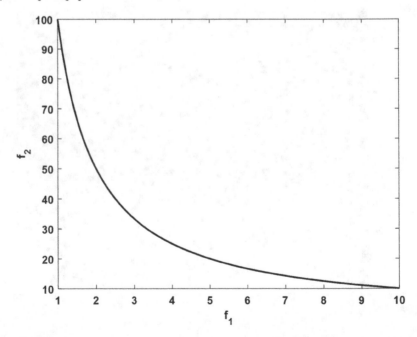

FIGURE 12.12
Pareto optimal front (after 50 generations; number of solutions in the Pareto optimal front = 407).

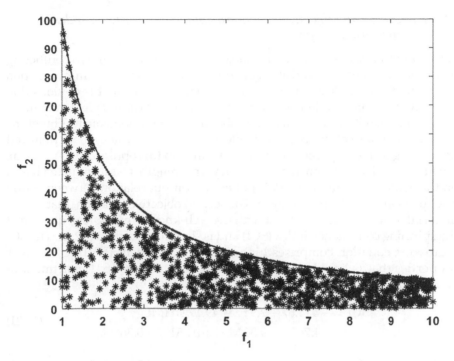

FIGURE 12.13
Objective space with Pareto optimal front.

12.4.1 Application of Evolutionary Algorithm Approach for Multiobjective Optimization

There have been a few reported works on the application of the GA for multiobjective optimization in the textile field. Sette and Lengenhove (2002) used the concept of multiobjective optimization in the spinning field to strike a balance between two conflicting factors, namely, price and quality of yarn. A similar concept was also reported to optimize shear and drape behaviors of woven fabrics (Skordos et al., 2002). Of late, the NSGA-II technique of multiobjective optimization was successfully applied by Ghosh et al. (2013), Majumdar et al. (2016), and Ghosh and Mal (2016). NSGA-II was used by Ghosh et al. (2013) for simultaneous maximization of cotton yarn strength and minimization of raw material quality. Majumdar et al. (2016) optimized two conflicting physiothermal properties (i.e., air permeability and thermal conductivity) of single jersey and rib-knitted fabrics with desired UV protection using NSGA-II. Ghosh and Mal (2016) applied an evolutionary algorithm optimizing three objectives, namely, air permeability, thermal conductivity, and UPF of single jersey knitted fabrics.

12.4.1.1 Using Two Objective Functions: Spinning Consistency Index and Yarn Strength

Ghosh et al. (2013) carried out multiobjective optimization of two conflicting objectives, namely, maximization of cotton yarn strength and minimization of raw material quality, using the evolutionary algorithm. It is obvious that an individual combination of cotton fiber parameters cannot yield maximum yarn strength with minimum cotton fiber quality simultaneously. Therefore, a set of optimal solutions is available from which a trade-off is required depending on the requirement of the consumer. In this optimization problem, the first objective function, that is, cotton yarn strength, was formulated based on the artificial neural network input equals output relation between cotton fiber properties and yarn strength. The second objective function, that is, raw material quality, was formulated with the well-known regression equation of the spinning consistency index (SCI) and is shown in Equation 12.21. The SCI regression equation comprises high-volume instrument (HVI) tested fiber properties that is based on the crop study data of the U.S. Department of Agriculture:

$$SCI = -414.67 + 2.9\,FS + 49.17\,UHML$$
$$+ 4.74\,UI - 9.32\,MIC + 0.65\,Rd + 0.36(+b) \tag{12.21}$$

The objective functions are shown in Equations 12.22 and 12.23:

$$Objective\ 1 = Yarn\ strength \tag{12.22}$$

$$Objective\ 2 = \frac{1}{1 + SCI^2} \tag{12.23}$$

NSGA-II was started with 100 randomly generated initial populations ($2N$), and it ranked the individuals based on dominance. The fast nondominated sorting procedure found the nondomination frontiers (ranks) where individuals of a particular frontier set are nondominated by any solution. By using this procedure, all of the initial scattered solutions were grouped into four frontiers after nine generations. In the next step, the crowding distance was calculated for each individual by applying Equation 12.18. The crowding distance selection operator helped in distributing the solution uniformly to the frontier rather than bunching up at several good points. Subsequently, steps 1–6 of NSGA-II were repeated, and the Pareto optimal front was obtained at the end of 98 generations leading to the final set of solutions.

The Pareto optimal front for SCI and yarn strength is illustrated in Figure 12.14. The Pareto optimal front contains 100 nondominated solutions. As all of the solutions in the Pareto front are better than the others, at least in terms of one objective, any one of them is an acceptable solution. The choice of one solution over the other exclusively depends on the requirement of the end user.

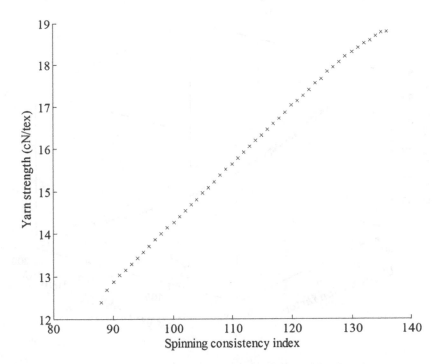

FIGURE 12.14

Pareto optimal front for SCI and yarn strength. (From Ghosh, A. et al. 2013. *Journal of Institute of Engineers India Series: E*, 94, 15–21.)

12.4.1.2 Using Three Objective Functions: Air Permeability, Thermal Conductivity, and Ultraviolet Protection Factor of Knitted Fabrics

Ghosh and Mal (2016) applied an elitist multiobjective evolutionary algorithm based on NSGA-II to simultaneously optimize three objectives functions, namely, air permeability, thermal conductivity, and UPF, of single jersey knitted fabric and obtained a set of Pareto optimal solutions. The quadratic regression equations mentioned in Table 12.1 for air permeability, thermal conductivity, and UPF were used for the optimization problem. The symbols X_1 and X_4 in the regression equations correspond to loop length and yarn count, respectively, which are the controllable factors for air permeability, thermal conductivity, and UPF in this problem. NSGA-II was started with 500 randomly generated initial populations ($2N$), and it ranked the individuals based on dominance. The fast nondominated sorting procedure found the nondomination frontiers (ranks) where individuals of a particular frontier set are nondominated by any solution. By using this procedure, the initial scattered solutions were grouped into four frontiers. In the next step, the crowding distance was calculated for each individual by applying Equation 12.18. The crowding distance selection operator helped in distributing the solution uniformly to the frontier rather

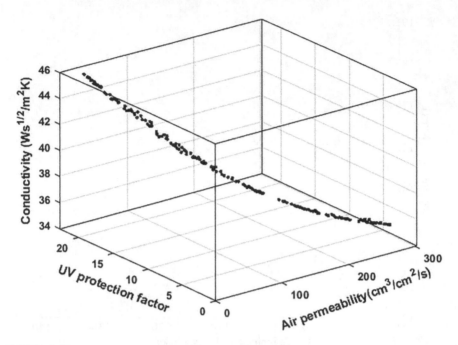

FIGURE 12.15

Pareto optimal front for air permeability, thermal conductivity, and UV protection factor. (From Ghosh, A. and Mal, P. 2016. Engineering design of knitted fabrics for optimal thermophysiological comfort as well as UV protection. *The Fiber Society 2016 Fall Meeting and Technical Conference*, New York, NY.)

than bunching up at several good points. Subsequently, steps 1–6 of NSGA-II were repeated, and the Pareto optimal front was obtained at the end of 200 generations leading to a final set of solutions.

Figure 12.15 shows the Pareto optimal front which helps in obtaining the effective controllable factors for engineering design of knitted fabrics with optimal thermophysiological comfort as well as UV protection. As all solutions in the Pareto front are the optimal solutions, the choice of a particular optimal solution over the others exclusively depends on the requirement of the end user.

12.5 MATLAB® Coding

MATLAB® Coding for Solving Single Goal Optimization Problem Given in Example 12.1

```
clc
clear all
```

```
close all
format short g
%
% Minimize d_m
% Subject to
% 3×1+ 2×2 <= 1400
% 5×1+10×2 <= 2800
% 40×1+30×2 <= 16800
% 50×1+40×2+d_m-d_p=21500
% x1, x2,d_m,d_p >= 0

f=[0 0 1 0];
b=[1400 2800 16800];
A=[3 2 0 0;5 10 0 0;40 30 0 0];
lb=[0 0 0 0];
ub=[];
Aeq=[50 40 1 -1];
beq=[21500];
[x fn_value]=linprog(f,A,b,Aeq,beq,lb,ub)

x1=x(1);
x2=x(2);
d_m=x(3);
d_p=x(4);

Obj_fn= d_m
Cons_1=3*x1 + 2*x2
Cons_2=5*x1 + 10*x2
Cons_3=40*x1 + 30*x2
Cons_4=50*x1 + 40*x2 + d_m - d_p
```

MATLAB® Coding for Solving Optimization Problem with Multiple Goals Given in Example 12.2

```
clc
clear all
close all
format short g

% Minimize d1_m+d2_p
% Subject to
% 2.5×1+ 2×2 <= 1400
% 5×1+10×2 <= 2800
% 30×1+30×2+d1_m-d1_p = 21500
% 40×1+30×2+d2_m-d2_p = 16800
% x1,x2,d1_m,d1_p,d2_m,d2_p <= 0

f=[0 0 1 0 0 1];
b=[1400 2800];
A=[2.5 2 0 0 0 0;5 10 0 0 0 0];
```

```
lb=[0 0 0 0 0 0];
ub=[];
Aeq=[30 30 1 -1 0 0;40 30 0 0 1 -1];
beq=[21500 16800]';
[x fn_value]=linprog(f,A,b,Aeq,beq,lb,ub)

x1=x(1);
x2=x(2);
d1_m=x(3);
d1_p=x(4);
d2_m=x(5);
d2_p=x(6);

Obj_fn= d1_m+d2_p
Cons_1=2.5*x1 + 2*x2
Cons_2=5*x1 + 10*x2
Cons_3=30*x1 + 30*x2 + d1_m - d1_p
Cons_4=40*x1 + 30*x2 + d2_m - d2_p
```

MATLAB® Coding for Solving the Optimization Problem (Preemptive Goal Programming) Given in Equation 12.9

```
clc
clear all
close all
format short g
%
% Minimize d1_m
% Subject to
% 2×1+ 3×2 <= 48
% 3×1+4×2 <= 64
% x2+d1_m-d1_p = 8
% x1+d2_m-d2_p = 6
% x1+x2+d3_m-d3_p = 15
% x1, x2,d1_m,d1_p,d2_m,d2_p,d3_m,d3_p >= 0

f=[0 0 1 0 0 0 0 0];
b=[48 64];
A=[2 3 0 0 0 0 0 0;3 4 0 0 0 0 0 0];
lb=[0 0 0 0 0 0 0 0];
ub=[];
Aeq=[0 1 1 -1 0 0 0 0;1 0 0 0 1 -1 0 0;1 1 0 0 0 0 1 -1];
beq=[8 6 15];
[x fn_value]=linprog(f,A,b,Aeq,beq,lb,ub)

x1=x(1);
x2=x(2);
d1_m=x(3);
d1_p=x(4);
```

```
d2_m=x(5);
d2_p=x(6);
d3_m=x(7);
d3_p=x(8);
Obj_fn= d1_m
Cons_1=2*x1 + 3*x2
Cons_2=3*x1 + 4*x2
Cons_3=x2 + d1_m - d1_p
Cons_4=x1 + d2_m - d2_p
Cons_5=x1 + x2 + d3_m - d3_p
```

MATLAB® Coding for Solving the Optimization Problem (Preemptive Goal Programming) Given in Equation 12.10

```
clc
clear all
close all
format short g

% Minimize d2_m
% Subject to
% 2×1+ 3×2 <= 48
% 3×1+4×2 <= 64
% x2+d1_m-d1_p = 8
% x1+d2_m-d2_p = 6
% x1+x2+d3_m-d3_p = 15
% d1_m=0
% x1, x2,d1_m,d1_p,d2_m,d2_p,d3_m,d3_p >= 0

f=[0 0 0 0 1 0 0 0];
b=[48 64];
A=[2 3 0 0 0 0 0 0;3 4 0 0 0 0 0 0];
lb=[0 0 0 0 0 0 0 0];
ub=[];
Aeq=[0 1 1 -1 0 0 0 0;1 0 0 0 1 -1 0 0;1 1 0 0 0 0 1 -1;0 0 1
0 0 0 0 0];
beq=[8 6 15 0];
[x fn_value]=linprog(f,A,b,Aeq,beq,lb,ub)

x1=x(1);
x2=x(2);
d1_m=x(3);
d1_p=x(4);
d2_m=x(5);
d2_p=x(6);
d3_m=x(7);
d3_p=x(8);
Obj_fn= d2_m
Cons_1=2*x1 + 3*x2
```

```
Cons_2=3*x1 + 4*x2
Cons_3=x2 + d1_m - d1_p
Cons_4=x1 + d2_m - d2_p
Cons_5=x1 + x2 + d3_m - d3_p
```

MATLAB® Coding for Solving the Optimization Problem (Preemptive Goal Programming) Given in Equation 12.11

```
clc
clear all
close all
format short g

% Minimize d3_m
% Subject to
% 2×1+ 3×2 <= 48
% 3×1+4×2 <= 64
% x2+d1_m-d1_p = 8
% x1+d2_m-d2_p = 6
% x1+x2+d3_m-d3_p = 15
% d1_m=0
% d2_m=0
% x1, x2,d1_m,d1_p,d2_m,d2_p,d3_m,d3_p >= 0

f=[0 0 0 0 0 0 1 0];
b=[48 64];
A=[2 3 0 0 0 0 0 0;3 4 0 0 0 0 0 0];
lb=[0 0 0 0 0 0 0 0];
ub=[];
Aeq=[0 1 1 -1 0 0 0 0;1 0 0 0 1 -1 0 0;1 1 0 0 0 0 1 -1;0 0 1
0 0 0 0 0;0 0 0 0 1 0 0 0];
beq=[8 6 15 0 0];
[x fn_value]=linprog(f,A,b,Aeq,beq,lb,ub)

x1=x(1);
x2=x(2);
d1_m=x(3);
d1_p=x(4);
d2_m=x(5);
d2_p=x(6);
d3_m=x(7);
d3_p=x(8);
Obj_fn= d3_m
Cons_1=2*x1 + 3*x2
Cons_2=3*x1 + 4*x2
Cons_3=x2 + d1_m - d1_p
Cons_4=x1 + d2_m - d2_p
Cons_5=x1 + x2 + d3_m - d3_p
```

MATLAB® Coding for Multiobjective Optimization Using Desirability Approach

```
clc
close all
clear all
x0=[-0.1 -0.9];
A=[];
B=[];
Aeq=[];
Beq=[];
vlb=[-1 -1];
vub=[1 1];
nonlincon=[];
options=optimset('LargeScale','on');
x1=[6.6 7.4];
x2=[5 10];
[x,fval]=fmincon(@dfnc_target,x0,A,B,Aeq,Beq,vlb,vub,nonlincon,
options);
x_max=x
max_dfnc=-fval
Optimum_val_x1=((x(1)+1)*(max(x1)-min(x1)))/2+min(x1)
Optimum_val_x2=((x(2)+1)*(max(x2)-min(x2)))/2+min(x2)
UPF=10.052 - 1.6867* x(1) - 7.6867*x(2)+ 1.3169*x(1).*x(2)+
2.0483*x(2).^2
TC=38.85-0.78*x(1)-4.6*x(2)+1.16*x(2).^2
AP=(134.25 + 22.29*x(1) + 111.49*x(2)+9.60*x(1).*x(2)+
25.94*x(2).^2)
A_min=36.01;
A_max=303.57;
A_target=80;
T_min=34.63;
T_max=45.39;
T_target=40;
U_min=4.04;
U_max=22.8;
U_target=15;
A=A_min:0.01:A_target;
d1=(A-A_min)/(A_target-A_min);
figure,plot(A,d1)
xlabel('Air permeability (cm^3 / cm^2 / s)')
ylabel('Desirability')
hold on
A=A_target:0.01:A_max;
d1=(A_max-A)/(A_max-A_target);
plot(A,d1)
set(gcf,'color','w')
T=T_min:0.01:T_target;
d2=(T-T_min)/(T_target-T_min);
```

```
figure,plot(T,d2)
xlabel('Thermal conductivity (W/mK x 10^-^3)')
ylabel('Desirability');
hold on
T=T_target:0.01:T_max;
d2=(T_max-T)/(T_max-T_target);
plot(T,d2)
set(gcf,'color','w')
U=U_min:0.01:U_target;
d3=(U-U_min)/(U_target-U_min);
figure,plot(U,d3)
xlabel('UPF')
ylabel('Desirability')
hold on
U=U_target:0.01:U_max;
d3=(U_max-U)/(U_max-U_target);
plot(U,d3)
set(gcf,'color','w')


x1=-1:0.05:1;
x2=-1:0.05:1;
[X1,X2]=meshgrid(x1,x2);
[m,n]=size(X1);
 for i=1:m
   for j=1:n
   Z(i,j)=-dfnc_target([X1(i,j),X2(i,j)]);
   end
end
 figure,surf(X1,X2,Z)
 xlabel('Loop length \it(X_1)\rm')
ylabel('Yarn count \it(X_4)\rm')
zlabel('Overall desirability')
set(gcf,'color','w')
COLORBAR

function Y=dfnc_target(X)
A=134.25 + 22.29*X(1) + 111.49*X(2) +9.60*X(1).*X(2) +
25.94*X(2).^2;
T=38.85 - 0.78*X(1) - 4.60*X(2) + 1.16*X(2).^2;
U=10.05 - 1.69*X(1) - 7.69*X(2) + 1.32*X(1).*X(2) +
2.05*X(2).^2;
A_min=36.01;
A_max=303.57;
A_target=80;
T_min=34.63;
T_max=45.39;
T_target=40;
U_min=4.04;
U_max=22.8;
U_target=15;
```

```
if A<=A_target
   d1=(A-A_min)/(A_target-A_min);
elseif A_target<A
   d1=(A_max-A)/(A_max-A_target);
else d1=0;
end
if T<=T_target
   d2=(T-T_min)/(T_target-T_min);
elseif T_target<T
   d2=(T_max-T)/(T_max-T_target);
else d2=0;
end
if U<=U_target
   d3=(U-U_min)/(U_target-U_min);
elseif U_target<U
   d3=(U_max-U)/(U_max-U_target);
else d3=0;
end
Y=-((d1*d2*d3)^(1/3));
```

MATLAB® Coding for the NSGA-II Example

```
clc
clear all
close all
format short g
pop_size=1000;
gen=1;
maxgen=50;
pc=0.8;
pm=0.001;
vlb=[1 1];
vub=[10 100];
bits=[8 8];
range1=10-1;
range2=100-(1/10);
n=length(bits);
len_chrom=sum(bits);
old_gen=round(rand(pop_size,len_chrom));
for i=1:n
    accuracy(i)=(vub(i)-vlb(i))./(2^bits(i)-1);
end
%NSGA-II program starts
while gen<maxgen
 dec_val(1:pop_size,1:n)=0;
 m=1;
 for i=1:n
 cg=m;
   for j=bits(i)-1:-1:0
   dec_val(:,i)=dec_val(:,i)+(2^j*old_gen(:,cg));
```

```
 cg=cg+1;
 m=cg;
 end
end
for j=1:n
 for i=1:pop_size
 x(i,j)=vlb(j)+dec_val(i,j).*accuracy(j);
 end
end
for i=1:pop_size
f1(i)=x(i,1);
f2(i)=x(i,2)./x(i,1);
end
obj=[f1' f2'];
R1=[];
R2=[];
R3=[];
R4=[];
count1=0;
count2=0;
count3=0;
%Finding front1
 pop_size1=pop_size;
 s1=0;t1=0;
 [m1,n1]=size(obj);
 for i=1:m1
  for j=1:m1
   if i~=j
    if (obj(i,1)>obj(j,1)&& obj(i,2)>=obj(j,2))||(obj(i,1)>=
obj(j,1)&& obj(i,2)>obj(j,2))
    s1=s1+1;
    S1(:,s1)=[i, j];
    elseif (obj(i,1)<obj(j,1)&& obj(i,2)<=obj(j,2))||(obj(i,1)
<=obj(j,1)&& obj(i,2)<obj(j,2))
    t1=t1+1;
    N1(:,t1)=[i, j];
    end
   end
  end
 end
 if s1==0 && t1==0
ndom1=1:pop_size1;
rest1=[];
 for i=1:length(ndom1)
 R1(i)=1;
 end
R2=[];
R3=[];
R4=[];
else
```

```
NN1=N1(1,:);
 for i=1:pop_size1
 e1= find(NN1==i);
 count1(i)=length(e1);
 end
ndom1=find(count1==0);
rest1=find(count1);
 for i=1:length(ndom1)
 R1(i)=1;
 end
end
front1=[R1' ndom1'];
%Finding front2
pop_size2=length(rest1);
obj2=obj(rest1,:);
s2=0;t2=0;
[m2,n2]=size(obj2);
 for i=1:m2
  for j=1:m2
   if i~=j
    if (obj2(i,1)>obj2(j,1)&& obj2(i,2)>=obj2(j,2))||(obj2
(i,1)>=obj2(j,1)&& obj2(i,2)>obj2(j,2))
    s2=s2+1;
    S2(:,s2)=[i, j];
    elseif (obj2(i,1)<obj2(j,1)&& obj2(i,2)<=obj2(j,2))||(ob
j2(i,1)<=obj2(j,1)&& obj2(i,2)<obj2(j,2))
    t2=t2+1;
    N2(:,t2)=[i, j];
     end
    end
   end
  end
  if s2==0 && t2==0
  ndom2=1:pop_size2;
  rest2=[];
   for i=1:length(ndom2)
   R2(i)=2;
   end
  R3=[];
  R4=[];
  ndom2=rest1(ndom2);
  else
  NN2=N2(1,:);
   for i=1:pop_size2
   e2= find(NN2==i);
   count2(i)=length(e2);
   end
  ndom2=find(count2==0);
  rest2=find(count2);
  ndom2=rest1(ndom2);
```

```
  rest2=rest1(rest2);
   for i=1:length(ndom2)
   R2(i)=2;
   end
  end
  front2=[R2' ndom2'];
  %Finding front3
  pop_size3=length(rest2);
  obj3=obj(rest2,:);
  s3=0;t3=0;
  [m3,n3]=size(obj3);
  for i=1:m3
   for j=1:m3
    if i~=j
     if (obj3(i,1)>obj3(j,1)&& obj3(i,2)>=obj3(j,2))||(obj3
(i,1)>=obj3(j,1)&& obj3(i,2)>obj3(j,2))
      s3=s3+1;
      S3(:,s3)=[i, j];
      elseif (obj3(i,1)<obj3(j,1)&& obj3(i,2)<=obj3(j,2))||
(obj3(i,1)<=obj3(j,1)&& obj3(i,2)<obj3(j,2))
      t3=t3+1;
      N3(:,t3)=[i, j];
      end
     end
    end
  end
  if s3==0 && t3==0
  ndom3=1:pop_size3;
  rest3=[];
   for i=1:length(ndom3)
   R3(i)=3;
   end
  R4=[];
  ndom3=rest2(ndom3);
  else
  NN3=N3(1,:);
   for i=1:pop_size3
   e3= find(NN3==i);
   count3(i)=length(e3);
   end
  ndom3=find(count3==0);
  rest3=find(count3);
  ndom3=rest2(ndom3);
  rest3=rest2(rest3);
   for i=1:length(ndom3)
   R3(i)=3;
   end
  end
  front3=[R3' ndom3'];
%Finding front4
```

```
  for i=1:length(rest3)
  R4(i)=4;
  end
  front4=[R4' rest3'];
  max_count=pop_size/2;
%crowding distance front1
  x1=f1(ndom1);
  [X1,i1]=sort(x1);
  d1_front1=zeros(length(ndom1),1);
  for i=1:length(ndom1)
   if i==1||i==length(ndom1)
   d1_front1(i)=Inf;
   else
   d1_front1(i)=d1_front1(i)+(X1(i+1)-X1(i-1))/range1;
   end
  end
  [c1_front1,t1_front1]=sort(i1);
  D1_front1=d1_front1(t1_front1);
  y1=f2(ndom1);
  [Y1,j1]=sort(y1);
  d2_front1=zeros(length(ndom1),1);
  for i=1:length(ndom1)
   if i==1||i==length(ndom1)
   d2_front1(i)=Inf;
   else
   d2_front1(i)=d2_front1(i)+(Y1(i+1)-Y1(i-1))/range2;
   end
  end
  [c2_front1,t2_front1]=sort(j1);
  D2_front1=d2_front1(t2_front1);
  D_front1=D1_front1+D2_front1;
  [S_front1,I_front1]=sort(D_front1,'descend');
  Order_front1=[ndom1(I_front1)' R1' S_front1];
%crowding distance front2
  x2=f1(ndom2);
  [X2,i2]=sort(x2);
  d1_front2=zeros(length(ndom2),1);
  for i=1:length(ndom2)
   if i==1||i==length(ndom2)
   d1_front2(i)=Inf;
   else
   d1_front2(i)=d1_front2(i)+(X2(i+1)-X2(i-1))/range1;
   end
  end
   [c1_front2,t1_front2]=sort(i2);
   D1_front2=d1_front2(t1_front2);
   y2=f2(ndom2);
   [Y2,j2]=sort(y2);
   d2_front2=zeros(length(ndom2),1);
  for i=1:length(ndom2)
```

```
  if i==1||i==length(ndom2)
  d2_front2(i)=Inf;
  else
  d2_front2(i)=d2_front2(i)+(Y2(i+1)-Y2(i-1))/range2;
  end
 end
 [c2_front2,t2_front2]=sort(j2);
 D2_front2=d2_front2(t2_front2);
 D_front2=D1_front2+D2_front2;
 [S_front2,I_front2]=sort(D_front2,'descend');
 Order_front2=[ndom2(I_front2)' R2' S_front2];
%crowding distance front3
 x3=f1(ndom3);
 [X3,i3]=sort(x3);
 d1_front3=zeros(length(ndom3),1);
 for i=1:length(ndom3)
  if i==1||i==length(ndom3)
  d1_front3(i)=Inf;
  else
  d1_front3(i)=d1_front3(i)+(X3(i+1)-X3(i-1))/range1
  end
 end
 [c1_front3,t1_front3]=sort(i3);
 D1_front3=d1_front3(t1_front3);
 y3=f2(ndom3);
 [Y3,j3]=sort(y3);
 d2_front3=zeros(length(ndom3),1);
 for i=1:length(ndom3)
  if i==1||i==length(ndom3)
  d2_front3(i)=Inf;
  else
  d2_front3(i)=d2_front3(i)+(Y3(i+1)-Y3(i-1))/range2;
  end
 end
 [c2_front3,t2_front3]=sort(j3);
 D2_front3=d2_front3(t2_front3);
 D_front3=D1_front3+D2_front3;
 [S_front3,I_front3]=sort(D_front3,'descend');
 Order_front3=[ndom3(I_front3)' R3' S_front3];
%crowding distance front4
 x4=f1(rest3);
 [X4,i4]=sort(x4);
 d1_front4=zeros(length(rest3),1);
 for i=1:length(rest3)
  if i==1||i==length(rest3)
  d1_front4(i)=Inf;
  else
  d1_front4(i)=d1_front4(i)+(X4(i+1)-X4(i-1))/range1;
  end
 end
```

```
 [c1_front4,t1_front4]=sort(i4);
 D1_front4=d1_front4(t1_front4);
 y4=f2(rest3);
 [Y4,j4]=sort(y4);
 d2_front4=zeros(length(rest3),1);
 for i=1:length(rest3)
  if i==1||i==length(rest3)
  d2_front4(i)=Inf;
  else
  d2_front4(i)=d2_front4(i)+(Y4(i+1)-Y4(i-1))/range2;
  end
 end
 [c2_front4,t2_front4]=sort(j4);
 D2_front4=d2_front4(t2_front4);
 D_front4=D1_front4+D2_front4;
 [S_front4,I_front4]=sort(D_front4,'descend');
 Order_front4=[rest3(I_front4)' R4' S_front4];
 Soln_Order=[Order_front1; Order_front2; Order_front3;
Order_front4];
 P=Soln_Order(1:max_count,1);
%Tournament selection
 Parent=Soln_Order(1:max_count,:);
 w=randperm(max_count);
 v=P(w);
 for i=1:max_count
  if i~=max_count
  T(i,:)=[v(i) v(i+1)];
  elseif i==max_count
  T(i,:)=[v(1) v(end)];
  end
 end
 u=randperm(max_count);
 Tournament=T(u,:);
 for i=1:max_count
  for j=1:max_count
   if Tournament(i,1)==Parent(j,1)
   R_1(i)=Parent(j,2);
   Crowd_1(i)=Parent(j,3);
   end
   if Tournament(i,2)==Parent(j,1)
   R_2(i)=Parent(j,2);
   Crowd_2(i)=Parent(j,3);
   end
  end
  if R_1(i)==R_2(i)
  [a,b]=sort([Crowd_1(i) Crowd_2(i)],'descend');
  Winner(i)=Tournament(i, b(1));
  else
  [c,g]=sort([R_1(i) R_2(i)]);
  Winner(i)=Tournament(i, g(1));
```

```
     end
   end
   Po=Parent(:,1);
   Qo=Winner';
   m_pool=Qo;
   h=randperm(pop_size/2);
   rand_m_pool=m_pool(h);
   Mating_pool=old_gen(rand_m_pool,:);
%Cross Over
   for i=1:2:pop_size/2
    if ceil(pop_size*rand/2)<=ceil(pop_size*pc/2)
     for j=1:len_chrom
      if rand<=0.5
      X_pool(i,j)=Mating_pool(i,j);
      X_pool(i+1,j)=Mating_pool(i+1,j);
      else
      X_pool(i,j)=Mating_pool(i+1,j);
      X_pool(i+1,j)=Mating_pool(i,j);
      end
     end
     G(i)='Y';
     G(i+1)='Y';
    else
    X_pool(i,:)=Mating_pool(i,:);
    X_pool(i+1,:)=Mating_pool(i+1,:);
    G(i)='N';
    G(i+1)='N';
    end
   end
   X_pool;
   Col_G=G';
%Mutation
   mu=0;
   for i=1:pop_size/2
    for j=1:len_chrom
     if ceil(pop_size*len_chrom*rand) <=ceil(pop_size*len_
     chrom*pm)
      if X_pool(i,j)==0
      mut_pool(i,j)=1;
      else
      mut_pool(i,j)=0;
      end
      mu=mu+1;
      [i j];
      else
      mut_pool(i,j)=X_pool(i,j);
     end
    end
   end
   X_pool;
```

```
mutation_occurance_counter=mu;
mut_pool;
scatter(obj(Order_front1(:,1),1),obj(Order_front1(:,1),2),'.k')
xlabel('f_{1}')
ylabel('f_{2}')
set(gcf,'color','w')
box on
pause(0.001)
old_gen=[old_gen(Po,:);mut_pool];
gen=gen+1;
ct=gen;
end
x1=x(Order_front1(:,1),1);
x2=x(Order_front1(:,1),2);
f1=obj(Order_front1(:,1),1);
f2=obj(Order_front1(:,1),2);
pareto_solutions=[x1 x2 f1 f2]
number_of_pareto_solutions=length(x1)
```

12.6 Summary

The concept of multiobjective optimization is introduced in this chapter. Both classical and ideal multiobjective optimization techniques, like GP, desirability function, and evolutionary algorithm, are discussed with suitable examples and coding in MATLAB® language. Published works on multiobjective optimization in textile manufacturing are also cited.

References

Asim, F., Mahmood, M. and Siddiqui, M. A. 2011. Optimization of process parameters for simultaneous fixation of reactive printing and crease resistant finishing using desirability function. *World Congress on Engineering and Computer Science, 2011*, San Francisco, CA, Vol. II.

Deb, K. 2001. *Multiobjective Optimization Using Evolutionary Algorithms*. Wiley, Chichester, UK.

Deb, K. 2010. *Optimization for Engineering Design*. PHI Learning, New Delhi, India.

Deb, K., Pratap, A., Agarwal, S. and Meyarivan, T. 2002. A fast and elitist multiobjective genetic algorithm: NSGA II. *IEEE Transaction on Evolutionary Computation*, 6(2), 181–197.

Ghosh, A., Das, S. and Banerjee, D. 2013. Multi-objective optimization of yarn quality and fibre quality using evolutionary algorithm. *Journal of Institute of Engineers India Series: E*, 94, 15–21.

Ghosh, A. and Mal, P. 2016. Engineering design of knitted fabrics for optimal thermo-physiological comfort as well as UV protection. *The Fiber Society 2016 Fall Meeting and Technical Conference*, New York, NY.

Ghosh, A., Mal, P., Majumdar, A. and Banerjee, D. 2016. Optimization of knitted fabric comfort and UV protection using desirability function. *Journal of Engineered Fibers and Fabrics*, 11(4), 20–28.

Goldberg, D. E. 1989. *Genetic Optimization for Search, Optimization and Machine Learning*, Addison-Wesley, Reading, MA.

Goldberg, D. E. and Deb, K. 1991. A comparative analysis of selection schemes used in genetic algorithms. *Foundations of Genetic Algorithms*, 1, 69–93.

Majumdar, A., Mal, P., Ghosh, A. and Banerjee, D. 2016. Multi-objective optimization of air permeability and thermal conductivity of knitted fabrics with desired ultraviolet protection. *Journal of the Textile Institute*, 108(1), 110–116.

Panneerselvam, R. 2002. *Operations Research*. Prentice Hall, New Delhi, India.

Sette, S. and Lengenhove, L. V. 2002. Optimizing the fibre-to-yarn production process: Finding a blend of fibre qualities to create an optimal price/quality yarn. *Autex Research Journal*, 2, 57–63.

Skordos, A. A., Sutcliffe, M. P. F., Klintworth, J. and Adolfsson, P. 2002. Multi-objective optimisation of woven composite draping using genetic algorithms. *27th International Conference SAMPE EUROPE*, Paris, France.

Souid, H., Babay, A. and Sahnoun, M. 2012. Fabric quality optimization by using desirability function and neural networks. *International Journal of Computer Technology and Applications*, 3, 356–364.

Srinivas, N. and Deb, K. 1995. Multiobjective optimization using nondominated sorting in genetic algorithms. *Journal of Evolutionary Computations*, 2(3), 221–248.

Taieb, A. H. and Msahli, S. 2013. Optimization of the knitted fabric quality by using multicriteria phenomenon tools. *International Journal of Fiber and Textile Research*, 3, 66–77.

Vohra, N. D. 2010. *Quantitative Techniques in Management*. Tata McGraw-Hill, New Delhi, India.

Index